Muskism

Muskism

A Guide for the Perplexed

QUINN SLOBODIAN
and BEN TARNOFF

ALLEN LANE
an imprint of
PENGUIN BOOKS

ALLEN LANE

UK | USA | Canada | Ireland | Australia
India | New Zealand | South Africa

Allen Lane is part of the Penguin Random House group of companies
whose addresses can be found at global.penguinrandomhouse.com

Penguin Random House UK
One Embassy Gardens, 8 Viaduct Gardens, London SW11 7BW

penguin.co.uk

First published 2026
001

Copyright © Quinn Slobodian and Ben Tarnoff, 2026

The moral right of the authors has been asserted

Penguin Random House values and supports copyright.
Copyright fuels creativity, encourages diverse voices, promotes freedom
of expression and supports a vibrant culture. Thank you for purchasing
an authorized edition of this book and for respecting intellectual property
laws by not reproducing, scanning or distributing any part of it by any
means without permission. You are supporting authors and enabling
Penguin Random House to continue to publish books for everyone.
No part of this book may be used or reproduced in any manner for the
purpose of training artificial intelligence technologies or systems. In accordance
with Article 4(3) of the DSM Directive 2019/790, Penguin Random House
expressly reserves this work from the text and data mining exception.

Set in 12.8/16pt Dante MT Std
Typeset by Six Red Marbles UK, Thetford, Norfolk
Printed and bound in Great Britain by Clays Ltd, Elcograf S.p.A.

The authorized representative in the EEA is Penguin Random House Ireland,
Morrison Chambers, 32 Nassau Street, Dublin D02 YH68

A CIP catalogue record for this book is available from the British Library

ISBN: 978–0–241–80511–4

Penguin Random House is committed to a sustainable future
for our business, our readers and our planet. This book is made from
Forest Stewardship Council® certified paper.

Contents

Introduction: An Operating System for the Twentieth Century vii

PART ONE
Foundation

1. Fortress Futurism   3
2. The Superset   18
3. Sovereignty as a Service   32
4. Electric Autonomy   57

PART TWO
Cyborg

5. Attention Alchemy   84
6. Cybernetic Collectives   103
7. Godwin's Engine   121
8. State X   139

Conclusion: Four Futures for Muskism   156

Acknowledgements   169
Notes   171
Index   225

Introduction:
An Operating System for the Twenty-First Century

Everyone has an opinion about Elon Musk. He is a genius entrepreneur, launching humanity toward a science-fiction future. Or a ketamine-addled meme-lord, inflating bubbles and babbling about birth rates. Or, more recently, a cat's paw of the far right, his brain rotted by Twitter and dark prophecies of migrant invasion.

The verdicts differ but they share one thing: they treat Musk as an individual. Savior, clown, villain, addict. But good history looks past the singular psyche. When we—a historian and a tech writer—started discussing this book, we thought the more useful question is not *who is Musk?* but *what is Musk a symptom of?* What follows is our attempt at an answer, drawn from what Musk has publicly said and done, as documented in the citations that appear at the end of this book.

A century ago, Henry Ford wrote his best-selling memoir *My Life and Work*. Soon after, people coined a term: "Fordism." Out of one man came a new common sense. Fordism was more than cars rolling off assembly lines; it became shorthand for twentieth-century capitalism, built on the pairing of mass production with mass consumption.[1]

We treat Musk the same way. As others have suggested, he is not just a man but the avatar of a worldview: Muskism.[2] This is

INTRODUCTION

not his term—just as Ford never spoke of Fordism. If Fordism was the operating system of the twentieth century, we contend that Muskism offers a possible operating system for the twenty-first.

Like Fordism, Muskism is a modernizing project. But Fordism rewrote the social contract with a promise of rising living standards for all: cars in every garage, fridges in every kitchen, wages climbing with productivity. Muskism does not distribute rewards broadly. Its promise is sovereignty through technology.

Musk does not just sell cars, rockets, or satellites. He sells the fantasy that, in an increasingly unstable world, both states and individuals can fortify their self-reliance by plugging into *his* infrastructures. The paradox is that, in doing so, you become reliant on him. What is sold as *techno-sovereignty* is entry into Musk's walled garden, to which he holds the master key. Both the Pentagon and NASA depend on SpaceX; Starlink has become indispensable on the battlefield and in the wilderness; X and Grok are being woven into the state. Trying to unplug from Musk, you realize he owns the socket.

This techno-sovereignty is also selective. It offers autonomy for some and exclusion for others. Migrant hordes and their liberal enablers are vectors of a "woke mind virus" that needs to be traced, contained, and neutralized. Muskism sees the world as corrupted code. Empathy for one's fellow humans is an "exploit"—a vulnerability in our mental software—manipulated by bad actors to push the West toward "civilizational suicide."[3] "Suicidal empathy is like an autoimmune disease," Musk says, "the body attacks itself."[4]

If one face of Muskism is techno-sovereignty, the other is expulsion. Countermeasures include purged social networks, ideologically cleansed AI models and mass deportation of ethnic outsiders. The end goal is a purified community defined by cultural and genetic membership in a white, European West garrisoned by superior technology—a fortress to protect the best of humanity from the worst. The technologies of Muskism's walled garden will fortify the walls of the nation and the home. Harden your heart, harden your borders and debug the codebase. "If tolerance means the end of Western Civilization," he posted to his 225 million followers in 2025, "then we cannot be tolerant."[5]

What's the point of thinking about Muskism rather than Musk? For one thing, Muskism helps clarify Musk. Many still think of him as a libertarian who despises government. We think the reality is inverted: Musk has built his empire by fusing with the state. He repeatedly discusses his desire to colonize Mars, which he describes as his life's work. The logic of Muskism shows that Mars was never a serious exit plan—it is a bargaining chip, leverage for further techno-sovereigntist pursuits. Musk's online persona is similarly misunderstood. Critics see immaturity or malice; fans see relatability or authenticity. Both fail to see that, in Muskism, trolling is infrastructure. Every joke, every poll is a stress test of responsiveness: can he still move markets with a post? Can he tutor the algorithm and the underlying training data to push it further rightward? Can he simulate democracy through a reply-guy plebiscite? This is not play, but experiment. Wittingly or not, Musk is measuring and manipulating the elasticity of attention, the bandwidth of belief.

Muskism also envisions a less human future. Through automation, humans are purged from the productive process. Through social media, brain–computer interfaces, and artificial intelligence, humans are merged with the machine, forming what he calls a "cybernetic collective." The promise of sovereignty through technology acquires a cyborg form.

Musk is often seen as invincible. Yet the foundations of his kingdom are fragile. One of the less noticed resemblances

INTRODUCTION

between him and Ford is the extreme illiquidity of their personal wealth. Nearly all of Ford's fortune was in his Ford stock, which was privately held until almost a decade after his death. Musk's wealth is also almost entirely held in the stock of his own companies. As he put it in an interview, "if Tesla and SpaceX went bankrupt, I would go bankrupt too immediately."[6]

This is why Muskism depends on the perpetual expectation of pending technological breakthroughs, planetary salvation, or financial windfalls. To sustain his wealth, Musk must sustain belief in the exponential future growth of his companies. He learned the financial value of such fabulism in Silicon Valley. The bubble needs to stay inflated. Looking at his biography, we can see that Musk's outlook is often downstream of the business cycle. When credit is cheap, his rhetoric expands. When money is tight, he sees enemies everywhere.

We are living in an era when Muskism could thrive. Across the democratic world, people's trust in institutions is at historic lows.[7] The growth of anti-migrant sentiment has empowered the far right, which is enjoying its greatest resurgence since the 1930s—with Musk as its loudest mouthpiece. Donald Trump is scuttling the liberal international order abroad while assailing the American constitutional order at home. The intensifying rivalry between the United States and China, along with the Russian invasion of Ukraine, has made for a more fragmented, paranoid, and militarized world. Israel's genocide in Gaza, conducted with full bipartisan support from the United States, has shredded the last pretense of international law.

Muskism is well suited to these developments. Its promise of sovereignty through technology is attuned to the politics of a deglobalizing world, where states increasingly value independence over integration. Its offer of autonomy for some and exclusion for others is aligned with the new anti-humanitarianism, under which certain populations are marked for banishment and death. Its fascination with cyborgs accords with the techno-maximalism of a political and business elite that finds every occasion to digitize our lives more deeply, most recently under the guise of artificial intelligence.

To say that Muskism is worth taking seriously is not to say that its success is guaranteed. But the institutional breakdown of our era offers an opening. At some point, society will stabilize on a new basis. Muskism could provide the foundation.

To be clear, we are not making the argument that a coherent set of beliefs has guided Musk's choices over the years. He is not a systematic thinker, nor someone guided by a fixed ideology. We are interested in not only what he says but also in what he does, and in the historical forces that have shaped those actions. Muskism can be found in the feedback loop between man and moment. To understand the world that Musk aims to build, we have to understand the worlds that built Musk.

This is what our book sets out to do. We tell the story of Musk's defining moments, from apartheid South Africa to meme coins, from the launch pad to the doomscroll. We explore how his improvisations, magnified by financial cycles and geopolitical crises, crystallized into Muskism. Our book is a guide for those perplexed not just by Musk but by the historical conjuncture we find ourselves in.

PART ONE

Foundation

Elon Musk has long cited Isaac Asimov's *Foundation* novels as a formative influence.[1] The series, widely considered a science-fiction classic, is set far in the future, in the waning days of the Galactic Empire, as a genius mathematician named Hari Seldon establishes an organization called the Foundation. The purpose of the Foundation is to preserve and advance human knowledge during the dark age that follows imperial collapse. Through a series of crises, the Foundation gradually expands its influence and grows into a new Galactic Empire.

Asimov was interested in how history moves from one epoch to the next. (Edward Gibbon's *History of the Decline and Fall of the Roman Empire* served as his initial inspiration.[2]) In *Foundation*, he presents a model of historical motion where eras do not follow one another in an orderly sequence, like a line of railroad cars. Rather, the seeds of the new era germinate inside the preceding one. The Foundation is created in the first Empire to serve as a nucleus for the second.

There is a similar story to be told about the emergence of Muskism over the past half-century. It arose from specific historical conditions while pointing beyond them toward something new.

The first part of this book tracks the evolution of Muskism through four formative periods of Musk's life. Chapter 1 examines his childhood in apartheid South Africa in the 1970s and 1980s. Chapter 2 takes up his early career in Silicon Valley during the dot-com boom of the 1990s, where he made his first fortune. Chapter 3 looks at SpaceX, which he founded in 2002 after cashing out of the internet economy and moving to Los Angeles. Finally, Chapter 4 is devoted to Tesla, which he took over as CEO in 2008 after having been an investor for years.

Together, these four chapters introduce the pillars of Muskism. They show how Muskism's promise of sovereignty through technology took shape within the most important political, economic, and cultural developments of the last fifty years. Like the Foundation, Muskism would be a new order, built in the shell of the old.

I
Fortress Futurism

"Let us wake up to the physical possibilities of a greater living," Elon Musk's maternal grandfather wrote in 1940 in the journal of Technocracy Incorporated.[1] Joshua Haldeman was a chiropractor and former bronc rider in Saskatchewan—and an active member in a movement that wanted to kill off capitalism and democracy at one stroke. Founded by the American engineer Howard Scott in 1933, Technocracy envisioned a society run under the dictatorship of engineers, who would allocate resources according to scientific principles. Currency would be replaced by stable units of energy called "ergs." This approach, combined with widespread automation, would ensure a life of greater leisure for all. Only those between twenty-five and forty-five would have to work, and only four hours a day, four days a week. "In Technocracy we see science replacing an economy of scarcity with an era of abundance," proclaimed Scott.[2]

Among believers, devotion to Technocracy was total. Members referred to themselves by numbers. (Haldeman's was 10450-1.)[3] They wore matching grey overcoats and drove convoys of matching grey Pontiac sedans. They even had a special salute. "The new culture of Technocratic man will be full of the mood of mastery," wrote Technocracy sympathizer E. Merrill Root, who would go on to help start the *National Review*: "The old culture is the wail of the impotent; the new culture will be the poetry of the potent."[4]

This future would require a new order: a superstate or "Technate" that would stretch from Greenland to the Galapagos. Only at such a scale could it command the resources for self-sufficiency. Economic independence would also shield the Technate from the corrosive entanglements of the global financial system. Haldeman showed a special fondness for this aspect of Technocracy. "Here on this Continent we have everything that is needed to provide us with certainty and security," he wrote in a 1940 article for the group's magazine. "Here, with a Technological Control in this Technological Age, the North American people could withstand the attack of the rest of the world combined."[5]

After the Canadian government banned Technocracy in 1940, Haldeman was briefly arrested. When his other political experiments proved uninspiring, he emigrated with his family—including Elon's mother, Maye, a convertible, and an airplane for backcountry adventures—to South Africa in 1950.[6] The timing was significant. The family arrived only two years after the National Party came to power and introduced a policy of "apartheid," or racial separatism. In fact, this was part of the draw for Haldeman. "Instead of the Government's attitude keeping me out of South Africa, it had precisely the opposite effect—it encouraged me to come and settle here," Haldeman told the pro-apartheid newspaper *Die Transvaler* soon after arriving.[7] To the regime's supporters, South Africa was a citadel of white civilization at the end of a Black continent, ringed by enemies. Later, in a self-published anti-Semitic screed, Haldeman denounced international opposition to apartheid as proof of a Jewish "conspiracy to establish a world dictatorship."[8]

But South Africa was not only a white-supremacist state. It

was also a modernizing one. Apartheid, the South African intellectual W. A. de Klerk argued, was "an attempt to remake a society in the total vision of a socio-political ideal."[9] The regime, as the historians Paul Edwards and Gabrielle Hecht remind us, pursued "a technopolitical project" that commanded the loyalty of those who subscribed to its prime directive: the construction of a utopia founded on racial segregation.[10]

Apartheid's architects saw themselves as futurists. They embraced technology, which they hoped would help them harden racial inequalities of an enduring and old-fashioned sort. They found support from international firms like IBM, Ford, and Toyota, which sold them technology to bolster their sovereignty. Mainframes were used to count, track, and reallocate Black laborers; blueprints for auto factories were used to create homegrown industry; nuclear weapons gave the regime the ultimate defense against its enemies. Singapore's first leader, Lee Kuan Yew, described turning his nation into a poisonous shrimp: indigestible to the predators around it.[11] South Africa, facing down the twin threats of *rooi gevaar* (red danger) and *swart gevaar* (black danger)—Communism and Black nationalism—played the same game. Only by adopting modern techniques and technologies could the state ensure its survival.

Though often dismissed as an anachronism, South Africa was in some respects—like its partner state Israel—a precursor to our own time. Its model of militarized, modernizing isolation fits more comfortably in today's world of export controls, trade wars, rearmament, and reshoring. Despite Musk's conflicted relationship with his country of origin, apartheid South Africa was the cradle of Muskism. It taught the lesson of fortress futurism: the belief that technology can strengthen self-reliance in a hostile world.

FOUNDATION

City of the future

Musk was born in 1971 in Pretoria. His hometown was a showcase of fortress futurism. As the regime's administrative capital, it housed the executive offices where apartheid was planned, and the security services that kept it in place. The architecture of its government buildings expressed the aspirations of apartheid's rulers. They were done in the sleek modernist minimalism of the International Style, with pilotis, podiums, and curtain walls, made of concrete, steel, and glass. The Pretoria of Musk's youth gleamed like a gearbox, as cutting-edge in appearance as Le Corbusier's capitals in Brasilia or Chandigarh.[12]

Close to the gearbox, at the city's edge, stood a cluster of brutalist structures. This was South Africa's main nuclear research site. It was called Pelindaba, a Zulu phrase that means "the end of the story."[13] Working with uranium gleaned from gold mining, and drawing on expertise from American and Israeli scientists, South Africa first developed nuclear power and then nuclear weapons. The regime had its first operational nuclear warhead by 1982.

The rest of Pretoria set apartheid in stone. The Black population was pushed into segregated townships on the outskirts of town. Meanwhile, the white elite clustered in suburbs like Waterkloof, where Musk lived during high school, perched on the ridge south of the city. These were enclaves of swimming pools and green yards, with quick roads into the ministries and military installations of the city center. They were not yet the gated communities they would become by the 1990s, but they didn't have to be. The entire state ensured their safety.

Dotting the city's perimeter were factories and warehouses.

These were sites of industrialization where South Africa hoped to build domestically the things it had previously imported. An interlocking system strove toward economic self-sufficiency. Pretoria West produced steel, ammunition, rubber, and plastic. To the east, Ford built a factory in Silverton in 1967. To the north-west, BMW constructed its first plant outside of Europe in Rosslyn in 1968.[14]

The car factories were built between the white city and the Black townships so that Black labor could be kept close at hand, tapped when needed and cast aside when not. Whereas the basic principle of Fordism was mass production plus mass consumption, South Africa's variant preserved mass consumption as a privilege for the white minority. In the 1980s, the political scientist Stephen Gelb gave the South African system a name: racial Fordism.[15]

Another key technology of productivity, security, and population management for the South African state was the computer. A Department of Planning established in 1965 envisioned the nation as a factory floor organized along racial Fordist lines.[16] Bureaucrats cleared so-called "black spots" from white areas through the forcible resettlement of Black communities—displacing three and a half million people by the early 1980s—while directing streams of cheap Black labor toward new industrial sites.[17]

Carrying out this vast project of social engineering required the accumulation of immense amounts of personal data. South Africa was what one historian calls a "biometric state."[18] Computers played a critical role in enabling the government to store and process the information required to implement apartheid, such as in the "Book of Life" identification system that used IBM mainframes to record the racial classification of every citizen. Anti-apartheid activists quickly grasped the dystopian potential of such tools. A 1982 pamphlet titled *Automating Apartheid* imagined a near-future scenario in which an operator asks a computer:

> Give me the names and addresses of all blacks on Victoria Street. Include pass numbers and fingerprints . . . the computer flashes the requested data onto a screen . . . At the same time, the information is electronically transmitted to the police . . . Imported computer technology makes this type of operation simple.[19]

Although the state's actual capacities fell short of such fears, the specter of techno-authoritarianism was real. From the outset, apartheid was a data-driven project—a reactionary technocracy that saw society as something to be optimized. Officials understood that computers could sharpen this project's precision. Yet digital tools were not only in the hands of the regime. Anti-apartheid groups also adapted them, using the Commodore 64—an early personal computer launched in 1982—to send encrypted messages to militants abroad.[20] Technology hardened the state's armor but could also make cracks in it.

The world in a box

Another person for whom the computer was a portal to the larger world was a young Elon Musk. A shy, nerdy child, Musk was alienated by the machismo that dominated white South African society. It was "not an intellectual culture," his father Errol recalled.[21] At school, Musk was bullied relentlessly. At home, he read constantly. When he ran out of books at the library, he later told a biographer, he devoured the entire

Encyclopaedia Britannica.[22] Musk belonged to the English-speaking minority of the white citizenry, which tended to be critical of apartheid but not necessarily antiracist in its outlook. Apartheid was the project of the Afrikaners, descendants of the Dutch settlers ("Boers") who had once fought the British for control of southern Africa and now made up the majority of the country's white minority. As the leaders of the ruling National Party, the Afrikaners were the most politically powerful faction of the population. "White, English-speaking South Africans such as Musk's family benefited from apartheid's racial hierarchy but lived mostly separate lives from the ruling Afrikaners," observes South African journalist Rachel Savage.[23]

Even so, anglophone whites did sometimes go into politics. Elon's father spent eleven years on Pretoria City's Council, representing the suburb of Sunnyside from 1972 to 1983.[24] In 1980, he joined the Progressive Federal Party, a predominantly white group of English-speaking reformists who opposed apartheid, but left after only three years because he disagreed with the party's antipathy to a new constitution that introduced a "tricameral" parliament, which gave limited representation to those citizens designated as "coloured" (people of mixed racial descent) and Indians while continuing to exclude the majority of the Black population.[25] Later, Errol would remember the apartheid era fondly. "There were no problems. People, Blacks and whites, got on very well with each other," he told the *Guardian* in 2025. "Everything worked. That's the reality. Of course, people don't want to hear that, but that's the truth."[26]

Musk attended Pretoria Boys High School, founded in 1901 as a finishing school for the British colonial elite. He and his classmates posed for photographs in striped boating blazers in imitation of Eton and Harrow, at a classic Anglo-South African remove from the more déclassé Afrikaners and the Black population.

"While the country as a whole was very much in flames and in turmoil, we were blissfully very safe in our little leafy suburbs, going about our very normal life," recalled one Pretoria Boys student who had been a year ahead of Musk.[27] At the same time, the school was not without its more liberal leanings, admitting its first Black student in 1981. Musk befriended the student's cousin and was one of the few white people to attend the boy's funeral after his death in a car crash in 1987.[28]

Most of all, Musk would have experienced the South Africa of the 1980s as unbearably provincial. This feeling would have only been sharpened by all the traveling he did as a child, made possible by Errol's wealth and frequent business trips. His father took him all over the world, from Hong Kong to the United States.[29] By contrast, his home country was a closed society, far from everywhere. "South Africa was like a prison for someone like Elon," his brother Kimbal recalled.[30]

But what was the way out? According to his own oft-repeated lore, Musk had an epiphany as an adolescent after reading Douglas Adams's 1979 sci-fi hit *The Hitchhiker's Guide to the Galaxy*.[31] The book's hero is Arthur Dent, a man rescued from his meaningless existence by an alien who takes him on an intergalactic adventure. The story is filled with intricately imagined futures. Most of it is unbelievably far-fetched, but there's one part that would have felt plausible to the young Musk: the Hitchhiker's Guide itself. Designed for travelers who need quick, useful information while roaming the universe, the Guide is a handheld electronic device. It "had about a hundred tiny flat press buttons and a screen about four inches square on which any one of a million 'pages' could be summoned at a moment's notice," writes Adams.[32]

In the early 1980s, Musk acquired just such a Guide for himself in the form of the Commodore VIC-20, one of the first affordable personal computers. Musk saw it for sale at the Sandton City Mall in Johannesburg and freaked out. "It was like, 'Whoa. Holy shit!'" he later told a biographer. He promptly "hounded" his father into buying it.[33] The VIC-20 didn't come with any pre-installed programs. But it did come with a programming language, BASIC, that you could use to write your own programs, as well as a *Programmer's Reference Guide* that taught you how. The book also advertised a range of applications for sale, from those for playing chess to those for composing songs, managing inventory, and performing "biorhythm charting." Musk stayed awake for three days straight learning to program the machine. "I just got super OCD," he later said.[34]

Alongside Apple and Radio Shack, Commodore was one of the companies that first introduced the paradigm of personal computing in the 1970s and 1980s. In doing so, it redefined programming. Instead of needing to go to a campus computer lab, novices like Musk could learn to program at home—and this enabled a more intimate relationship with the machine. The programmer merged their mind with the computer. They achieved a kind of cyborg consciousness, losing themselves in a digital flow state so complete that they no longer felt the presence of their body or the passage of time.

Musk used the Commodore to program his first product: a video game called *Blastar*, which involved shooting down an alien spacecraft. He sold it for $500 to a magazine, which published the source code for the benefit of other computer users.[35] The computer gave him the power to create virtual worlds. It also let him forge virtual connections with other parts of the physical world. Errol would later remember Musk showing him a "gray box with a red light on it"—a modem. "He said, 'With this thing . . . I can communicate with the computer at Oxford University,'" Errol recalled. "He was good at looking at the future."[36]

FOUNDATION

The mech

A paradox of the 1980s—the years of Musk's childhood—was that South Africa was becoming more repressive internally even as pinholes were opening up to the wider world. If computers offered one such aperture, another was provided by television. TV had only appeared in South Africa in 1976, one of the last industrialized countries to introduce the technology. It had held off because Hendrik Verwoerd, the prime minister commonly regarded as apartheid's chief architect, considered TV's destructive potential comparable to that of a nuclear bomb.[37] The wrong programming could corrupt the minds of the citizenry. "The government has to watch for any dangers to the people, both spiritual and physical," he cautioned.[38] (The approach had its admirers. In *National Review*, James Burnham wrote: "The absence of a 'native' liberation movement in South Africa is equivalent, very nearly, to the enforced absence of TV in South Africa.")[39]

Star Trek appeared on South African television in 1980 and the films screened in theaters.[40] Musk would later repeatedly cite the series as an inspiration. The ashes of the actor who played Scotty, James Doohan, were even, as one journalist put it, "accidentally re-cremated" in a failed SpaceX launch.[41] But there were other shows Musk must have watched as well, including *Battlestar Galactica* and *Buck Rogers in the 25th Century*, broadcast in South Africa in the early 1980s, and the animated series *Transformers* and *Robotech*, which aired in the decade's second half.[42] *Transformers* recounted, according to the South African TV listings, the "battle for supremacy" between two races of aliens, the good Autobots and the evil Decepticons.[43] They were both robots that transformed into vehicles—in this case, automobiles. Decades later, when Musk began to pour money into the development of humanoid machines at Tesla, he would pay homage to his childhood passions by naming the robot "Optimus" after Optimus Prime, the leader of the Autobots.

In *Robotech*, an American–Japanese production first released in 1985 and aired in South Africa the following year, an alien spacecraft crashes into Earth. The global community, previously at war, bands together to study the downed vessel. Eventually, the spacecraft is rebuilt and launched—just as a giant alien warrior race called the Zentraedi appear in Earth's orbit. But humanity now has a powerful weapon on its side, retrieved from the original ship: a technology that enables fighter jets to transform into manned humanoid robots. These are known as "mechs"—short for *mecha*, a Japanese term derived from the English word "mechanical." Mechs are such a popular feature of Japanese anime and manga that there is an entire subgenre devoted to them.

To an engineering-minded child like Musk, the mech would have been a beautiful thing. His father Errol worked as a mechanical and electrical engineer, and Musk himself showed an early aptitude for math and science. The mech was an engineering marvel. It integrated its human pilot so completely that man and machine were merged into a single entity. In *Robotech*, the mech even had the ability to act quasi-autonomously, without direct guidance from the pilot.

Robotech was a hit in South Africa. Action figures and branded Band-Aids were advertised in the newspaper, and a powdered orange drink printed a large ad commanding "secret *Robotech* operatives" to buy their product, with the teenage hero looking out from under his futuristic helmet.[44] Shows like *Transformers* and *Robotech* would feel especially resonant for white South African viewers in the 1980s. Their plotlines dramatized the country's predicament. Like Godzilla, the mech was a late echo of Japan's defining modern trauma: the nuclear weapons that destroyed Nagasaki and Hiroshima in 1945. Godzilla is a monster produced by nuclear radiation; the mech is another mutant of modernity.

Godzilla is both a destroyer and a savior. In some films, he threatens humanity; in others, he serves as Earth's protector. The mech embodies the same duality. It is an advanced technology possessed by the enemy that can be turned against the enemy. The connection to nuclear weapons could not be closer. The leaders of apartheid believed they faced an alien force that represented an apocalyptic threat. Only by constructing an apocalyptic power of their own could they safeguard their future. One did not have to squint much to see the Pretoria of the 1980s in the Autobot City of *Transformers* and the Macross City of *Robotech*. These were zones of fortress futurism: hardened, imperiled, mechanized.

For Musk, however, the meaning of the mech would run deeper. He would grow obsessed with the idea of becoming one with the machine. Later he would describe his company Neuralink as an attempt to create "a cyborg body that is incredibly capable" by means of brain implants.[45] In 2018, Musk tweeted to his more than twenty million followers, "It is time to create a mecha."[46]

For all its pretensions to technical prowess, South Africa had no mechs. One of its security solutions was to hurl young, poorly trained conscripts at the conflict on the northern border. Beginning in the 1960s, fighting had broken out between the South African Defence Force (SADF) and anticolonial guerrillas from present-day Namibia—and significantly intensified in the 1980s as Angolan troops entered the fray in greater numbers, with the help of the Cubans and the Soviets. Musk was destined to be such a conscript. By the age of seventeen, even upper-class white boys like him knew they would receive their call-up letter for mandatory service.

"I have been thrown into hell" is how the white South African novelist André Carl van der Merwe described his military experience, "herded into the Defence Force, into the abattoir of its border war like an animal to slaughter, with no say over my own destiny. Forced to kill people I don't know, for a cause I don't believe in."[47] Campaigns against conscription sprang up in opposition, denouncing "legalized murder."[48] "Free us from the Call-up," they demanded.[49] "He doesn't look like a terrorist," read another poster with a man in fatigues looking down at the legs of a dead body.[50]

Musk would later express a similar repugnance. "Spending two years suppressing black people didn't seem to be a great use of time," he said in 2013.[51] And, even more pointedly, in 1999: "Who wants to serve in a fascist army?"[52] So, at the age of seventeen, Musk did what almost none of his fellow South Africans had the privilege and possibility to do: he boarded a plane and left. He had secured a Canadian passport through his mother, a Canadian citizen, and his parents provided thousands of dollars in financial support.[53] He took a one-way plane ticket to Montreal, although he hoped to end up in California. He wanted to be "where the cutting edge of technology was," he later reminisced—and that meant Silicon Valley.[54]

By the time Musk left South Africa, fortress futurism had no future. The economy was in shambles. The achievements of the anti-apartheid struggle, combined with rising international pressure, pushed the regime to the breaking point. It made sense that Musk saw his future elsewhere. But many who leave home bring part of what they are abandoning with them.

One of the characteristics of South African life for white people was a blithe dependency on a vast and largely faceless class of laborers. Even for those who developed personal relationships with domestic servants, including nannies or maids, the Black population was mostly seen at the periphery of one's vision. Waiting at bus stops, in the long lines that you breezed past into government buildings or at border crossings. Floating above an indistinct but chromatically marked caste of fellow humans was part of the patrimony of whiteness. To the white population, they were "non-player characters" (NPCs), a concept from video games that would later furnish Musk with one of his favorite terms of abuse.

Apartheid modernism was the construction of a launchpad to a better future, one where, by definition, not all passengers would be brought on board. It was a star base of forty million, built for four million. It was founded on a logic of leverage, using human bodies as launch fuel for the next stage of leaving orbit. It held fast to the dream of omniscience and total control, to the belief that a territory could be treated like a factory. Its hostility to the liberal international order meant a commitment to sovereignty and hierarchy, to hard borders and white supremacy, to be secured through technology and technocracy. As the South African journalist Eve Fairbanks notes, apartheid taught a dual lesson: that "certain classic power hierarchies ought to be maintained *and* that those maintaining them deserve to be seen as disruptors, even misunderstood victims."[55]

In time, Muskism would incorporate these themes. The residue was identified by South African researchers who saw echoes in the Cybertruck of the armored vehicle that put down insurrections in the townships.[56] By 2025, the echoes were easier to hear as Musk became an impassioned advocate for the rights of white South Africans and used his social media platform to amplify false claims of "white genocide."[57]

The Pretoria of Musk's childhood could be left but not abandoned. In 1989, he flew to Canada, never to return.

Apartheid South Africa came along like a spore in his luggage.

2

The Superset

What Elon Musk saw in the blinking lights of the modem of his Pretoria suburb were the glimmers of a new age. By the time he settled in Silicon Valley in 1995, the growth of computer networks had spurred radical new visions of the nation state undone. Perhaps the most famous came from John Perry Barlow, the essayist whose 1996 manifesto "A Declaration of the Independence of Cyberspace" admonished governments—those "weary giants of flesh and steel"—that they had no jurisdiction in "Cyberspace, the new home of Mind."[1] Others imagined the dissolution of countries into fractal maps of "cyberstates" and "sovereign individuals."[2]

Musk took a different lesson from the 1990s. Silicon Valley would teach him that the real opportunity didn't lie in escaping the state but grafting onto it—using its guarantees as scaffolding for private gain. He had first glimpsed the formula as a college student in Canada. In 1991, while interning at Scotiabank, he had tried to convince his boss to buy a batch of "Brady Bonds" at a discount. These were created by U.S. Treasury Secretary Nicholas Brady, who helped U.S. banks get bad Latin American loans off their balance sheets by repackaging them as securities that were collateralized by U.S. Treasury bonds. His boss didn't go for it, explaining that the bank had already taken big losses on Latin American debt. "I tried to tell them that's not the point,"

Musk later recalled. "The point is that it's fucking backed by Uncle Sam."[3]

The takeaway for Musk was clear: the government isn't something to be minimized or eliminated. Rather, it can be instrumentalized as a source of power and profit. The goal wasn't state exit but state symbiosis. Musk's first Silicon Valley ventures in the 1990s would run directly on this logic: use free GPS data from military satellites to put maps online; piggyback on the federally ensured stability of the U.S. financial system to put banks online; and, above all, ride the dot-com boom triggered by the privatization of the internet, a technology invented by the government. The Commodore that Musk had persuaded his father to buy in the 1980s was little more than a glorified toy. A decade later, programmers were rewiring the world. The computer had become a control unit.

If South Africa was the nursery of Muskism, Silicon Valley would be its primary school. Here, Musk encountered the tools and the rules that showed how ideas could be turned into companies, and companies into monopolies, under conditions sustained by the state. The internet would turn everything into code. Uncle Sam would finance this future, but capital would own it.

Building the infomall

Using a modem, Musk's computer in Pretoria could talk over a phone line to a computer in Oxford. For a boy sitting in apartheid South Africa, a country where information was tightly controlled, the ease with which data could traverse thousands of miles and dozens of borders would have been astonishing. The internet, as it grew rapidly in the 1990s, would facilitate such flows on a vast scale.

The bland term "globalization" doesn't do justice to the scope of the decade's changes. World exports of goods and services tripled; global capital flows quadrupled.[4] New undersea cables connected ninety-two countries, boosting bandwidth by 147 times.[5] Between 1990 and 1997, the percentage of US households with computers went from 15 to 35 percent, while the amount spent by the average household on computers and associated hardware more than tripled.[6] In 1990, hardly anybody used the internet; by 2000, nearly half of Americans were online.[7] The first website appeared in 1991; by 2000, there were 17 million of them.[8]

"Humanity was effectively becoming a superorganism," Musk later mused, "qualitatively different than what it had been before."[9] The metaphor was borrowed from Kevin Kelly, an influential writer on technology and the founding executive editor of *Wired* magazine. Launched in 1993, *Wired* became the most important interpreter of the decade's transformations, observing the dawn of the networked age from its offices in downtown San Francisco. The year before the magazine's first issue, Kelly gave a foretaste of its grandiose house style in his 1992 book *Out of Control: The New Biology of Machines, Social Systems, and the Economic World*, in which he talked about how the internet was turning people into a "single interconnected superorganism," creating a "distributed intelligence."[10]

The organic analogies of self-ordering and decentralization always tactfully hid the direct role of the state. Silicon Valley as a whole owed its existence to the federal government, which bankrolled the region's tech industry after World War Two because the military needed electronics.[11] Even as the sector migrated to civilian customers, the government retained a large footprint, especially as a generous funder of research. Most of the innovations that enriched Silicon Valley were funded by the public sector, largely through the Defense Advanced Research Projects Agency (DARPA), the Pentagon's R&D arm—including the internet.[12] By the time Musk made it to Silicon Valley in the mid-1990s, however, the network was about to be privatized. In April 1995, the National Science Foundation, a federal agency, relinquished control of the internet's core infrastructure to the private sector, demanding nothing in return.[13]

The residents of the old internet were mostly researchers and academics. As late as 1993, addresses ending in .com (an abbreviation for "commercial") comprised a mere 1.5 percent of websites. By 1996, they accounted for fully half of the total.[14] The speed of the shift alarmed some observers, among them science-fiction writer William Gibson. "The thing I love about [the internet] is that it's transnational, non-profit—it isn't owned by anyone," he said in 1994. If corporations took control, he feared it would become an "infomall" where every byte came from a corporate menu.[15]

His fears were quickly realized. Four months after the internet's privatization, the dot-com boom began. The first spark came on August 9, 1995, when a startup co-founded by Marc Andreessen—born the same year as Musk—went public. As a student at the University of Illinois, Andreessen had helped create Mosaic, the first popular web browser, with National Science Foundation funding.[16] He then moved to Silicon Valley to launch Netscape—which, at the time of its IPO, hadn't turned a profit and mostly gave away its software. No matter: its stock price doubled on day one, closing with a valuation of $2.3 billion.[17] The internet was no longer a research network. It was an asset class.

Venture capital had long powered Silicon Valley, and splashy IPOs were nothing new—Apple's in 1980 had been the biggest since Ford's in 1956. But the hype around the internet exceeded anything in Silicon Valley's history. It promised not just new products but new markets and modes of life. Falling interest rates and strong post-1993 growth made capital abundant, while online brokerages attracted hordes of retail investors. Millions of Americans piled into tech stocks, inflating a bubble sustained by a single, seductive premise: the internet would change everything.

Financial fabulism

Musk joined the dot-com rush, incorporating his first company three months after the Netscape IPO. Zip2 began as an online directory of Bay Area businesses, complete with driving directions. He founded the startup with his brother Kimbal, who had followed him to California, and their friend Greg Kouri. They stitched together the data for their website from two sources. First, they bought a database of local commercial listings. Second, they acquired digital maps from Navigation Technologies, which relied on a Pentagon-built constellation of twenty-four GPS satellites that had gone fully operational in April 1995, the same month the internet was privatized. Like the internet, GPS was a state creation. Unlike the internet, it remained under direct U.S. military control. Musk and his brother convinced Navigation Technologies to let them use the maps for free until the company turned a profit.[18] It was a textbook example of state symbiosis: Zip2 rested entirely on publicly funded infrastructure, acquired at zero cost.

Even so, the startup struggled with the same question that almost all dot-com companies faced in the 1990s: how exactly could you make money online? The internet wasn't designed for commerce. Monetizing it would take innovation. At Zip2, Musk tried pivoting to licensing software to newspapers, signing clients like Hearst, Knight-Ridder, and *The New York Times*. His true aim, though, was not to strengthen legacy institutions but to bypass them entirely.[19] "Disruption" entered the popular lexicon in 1997, when business school professor Clayton Christensen published *The Innovator's Dilemma*, a text that became scripture in Silicon Valley.[20] Its argument—that nimble upstarts could topple entrenched incumbents by exploiting new technologies—offered a moral justification for the takeover of existing industries.

In February 1999, Compaq acquired Zip2 for $307 million. Musk walked away with $22 million.[21] Zip2 was losing money when Compaq bought it. He had been rewarded—but not for building a profitable business. If Silicon Valley showcased the value of state symbiosis, it also underscored another theme that would become equally integral to Muskism: financial fabulism. Fabulism is a literary genre that mixes fantasy and realism. The dot-commers, likewise, told stories about the extraordinary transformations that the internet would bring to people's everyday lives—and, by doing so, raised billions to build the future they foresaw. In the past, the typical Silicon Valley executive was low-key and technical, known for their engineering prowess. The flashy style of a Steve Jobs was the exception, not the rule. But in the 1990s, as tech grew more financialized, founders were forced to become more charismatic. To secure a good valuation, you had to be able to inspire confidence. Science fiction in the mouth of the right entrepreneur could conjure capital from thin air.

In the 1990s, the value of a startup lay not in current earnings but in imagined future profits. What set Musk apart wasn't engineering skill or business acumen, but his unshakable belief in the future—and his talent for making others believe in it too. When raising money for Zip2, he had built a big case around a normal PC and wheeled it around to venture capitalists to make them think his software ran on a supercomputer. "The investors thought that was impressive," Kimbal recalled.[22]

Where the money is

In 1995, Bill Gates prophesied that the internet would bring about a "friction-free capitalism."[23] Musk's second internet company would be an attempt to fulfill this promise by fusing information and capital. Money, like media, was easily rendered in code. "The Internet is about real-time, two-way digital information transfer and that's what a payment is," he told a reporter. "Payments are low bandwidth and they are digital. This is a gigantic opportunity."[24]

Founded in March 1999, X.com promised to fold the entire financial system—savings, checking, loans, mortgages, credit cards, mutual funds, brokerage, and insurance—into a single website. It would "be the place where all the money is," Musk declared.[25] But his timing was off. The dot-com boom was coming to an end. As the bubble began deflating in March 2000, even the most confident disruptors were forced into defensive moves.

X.com's solution was to merge with a competitor called Confinity, co-founded by Peter Thiel. At Confinity, Thiel and his cofounders had developed a web-based service for person-to-person payments called PayPal. By the spring of 2000, paypal.com was one of the world's most frequented online destinations, with more monthly users than the website of *The New York Times* and twice as many as that of the NFL.[26] Six months after the merger between X.com and Confinity closed, a power struggle ensued, which ended with Musk's ouster. Thiel became CEO and rebranded the company to match the name of its most successful product: PayPal.[27]

In 1994, the *Wired* writer Gary Wolfe had observed that the reason the Mosaic web browser became such a hit was that it was "pleasurable."[28] Users could glide from one page to another with a click. PayPal brought that same sensation to online consumption. Writing in 2000, the journalist Ric Manning recounted the hassle of using money orders or checks to pay for eBay purchases. "Economists call that problem friction," he wrote, "something that prevents the marketplace from functioning smoothly." PayPal, he concluded, was "the perfect lubricant."[29]

Many of the figures who made up what became known as the PayPal Mafia would have known this problem firsthand. Thiel had spent part of his childhood in apartheid South Africa and the South African territory of Southwest Africa (now Namibia); David Sacks was born in Cape Town a year after Musk; Roelof Botha was the grandson of apartheid South Africa's last foreign minister. All of them had grown up in a place where neither capital nor information moved freely. PayPal appeared to offer the antidote: a technology for turning cash into data and circulating it at the speed of light. Speaking to a *New York Times* journalist in 2000, Sacks made a remarkably prescient prediction. "Walk down the street, a few blocks away from your favorite Starbucks, pull out your Web-connected cell phone, you get a Starbucks menu, click espresso, and it's sent. And you've not only ordered it, but you've paid and you can go pick it up."[30]

Digital counterrevolution

The 1990s were a paradoxical time. On the one hand, the internet's growth stirred an outpouring of techno-utopianism. In the first issue of *Wired* in 1993, co-founder Louis Rossetto likened the "digital revolution" to the discovery of fire.[31] Rosetto was an offbeat libertarian whose itinerant resume included a three-week stint in South Africa in 1985. He adored it, especially the Bantustans, those impoverished, pseudo-independent territories created by the apartheid regime to give Black South Africans a simulacrum of self-rule.

For Rosetto, the quasi-sovereignty of the Bantustans was something to be celebrated: it liberated their residents from the control of traditional institutions.[32] Information technology, he believed, promised to do the same. This was a fringe view in the early 1990s but by decade's end it had gone mainstream. It came in different flavors, from the liberal variety embodied by Vice President Al Gore to the conservative strain associated with House speaker Newt Gingrich. A common thread, however, was an antipathy to bureaucracy, defined as an impersonal, rule-based order. Digital technology, and the internet in particular, promised to render bureaucracy obsolete by fostering more responsive and flexible forms of organization.

Yet all this talk of technological empowerment masked a very different reality. A term had already been coined in the 1990s to describe it: the "digital divide."[33] That explosion of new undersea cables? Only one of them had made its way to sub-Saharan Africa by the end of the millennium.[34] The glorious Babel being built online? As late as 1998, 75 percent of websites were using English.[35] Most of the internet's host computers were in the U.S., meaning network traffic within Asia often had to make a detour across the Atlantic and back.[36] In a detail worthy of Douglas Adams, the assignment of the internet's addresses remained in the hands of a single man with a scraggly beard and long hair until 1998: Jon Postel.[37] Meanwhile, the commercialization of the internet was creating a new elite in Silicon Valley, a process that was only briefly interrupted by the dot-com crash.

"Far from delivering us into a high-tech Eden," the scholar Dan Schiller wrote in 1999, "cyberspace itself is being rapidly colonized by the familiar workings of the market system."[38] To its idealists, the internet promised openness and horizontality; in practice, it tended toward monopoly and homogeneity. By making it to America, Musk hadn't escaped the stark world of spatial inequalities that divided his luxe, jacaranda-lined Pretoria suburb from the tin-shack townships just miles away. He had only rediscovered them on a larger scale.

What if the digital revolution was in fact a counter-revolution? Apartheid South Africa and Silicon Valley were not as far apart as they might seem. Both put their faith in technology and technocratic thinking. Both were infatuated with the engineer and the engineering mindset. (The German economist Friedrich von Gottl-Ottlilienfeld, reflecting on Fordism in 1925, praised its "dictatorship of technical reason."[39] The phrase could apply just as easily to apartheid South Africa as to Silicon Valley.) Of course, the stated values of the two places were different. Apartheid's engineers forged a system of racial oppression; the techno-utopianism of the 1990s, by contrast, equated digitization with democratization. In practice, however, the internet reconfigured social inequality rather than eliminating it. If Silicon Valley spoke the language of freedom, such language was misleading. Underneath, one could find the principle of reactionary technocracy at work.

Silicon monarchies

The dot-com meltdown that began in March 2000 was swift and brutal. "It's rare to see an industry evaporate as quickly and completely," a CNN journalist remarked.[40] In retrospect, however, in the words of Peter Thiel, "the dream of the '90s turned out to be right."[41]

What Thiel admired most about the dot-com era was its ambition: that "people believed in going from 0 to 1."[42] Going from zero to one, as he explains in his 2014 book of that title, means a singular act of technological creation. A company that invents something radically new can "earn" a monopoly—which should be the aim of any business.[43] Competition and capitalism are opposites, Thiel provocatively argues, since capitalism is about the accumulation of capital and you can't accumulate capital as easily if you're competing. A monopoly, by contrast, "owns its market, so it can set its own prices."[44] Further, such firms are actually good for the world, since their dominance lets them "make the long-term plans" and "finance the ambitious research projects that firms locked in competition can't dream of."[45]

If extreme concentrations of power benefit humanity, they are also natural. For Thiel, "severely unequal distributions" are simply the "law of the universe."[46] Not everyone can be Einstein and Shakespeare: in business as in life, a select few outpace the rest. This also happens to be the logic of venture capital: since most startups fail, VCs can only afford to invest in those companies that have the potential to yield enormous returns, to offset all of the losses in their portfolios. In Thiel's view, firms should be structured to reflect and reinforce the world's fundamental inequality. The most innovative companies "resemble feudal monarchies," with the founder as the king.[47] Only a suitably empowered monarch-monopolist "can make authoritative decisions, inspire strong personal loyalty, and plan ahead for decades."[48]

Thiel has always presented himself as a contrarian. But the value of *Zero to One*—for which Musk supplied an admiring blurb—lay in how sharply the book summarized the conventional wisdom of Silicon Valley as it consolidated in the dot-com era. Lose money for years in pursuit of growth. Make a market not a product. Don't worry about revenue until you've reached hyperscale, then use your monopoly position to extract superprofits. Idolize your founder as a princely visionary. (The cover of the February 19th, 1996 edition of *Time* showed a 24-year-old Marc Andreessen sitting barefoot on a golden throne.)

These ideas crystallized into a doctrine that would become central to Muskism: entrepreneurship as combat, with conquest as the goal. "When it comes down to it, for him, business is war," observed one of his X.com cofounders.[49] Musk loved strategy video games like *Civilization*, where players guide nations from infancy to supremacy—or suffer extinction. Startups became their real-world equivalents: small polities striving for supremacy in an unforgiving environment. As Thiel later put it, "A startup is the largest endeavor over which you can have definite mastery."[50] Muskism absorbed this creed wholesale: the founder as commander, the market as battlefield. Years later, Musk would change his official title at Tesla from CEO to Technoking.

But there was an important difference between Musk and Thiel. Thiel believed the internet would help bring about the full-scale privatization of sovereignty. One of his favorite books was *The Sovereign Individual* by James Dale Davidson and William Rees-Mogg, published in 1997, which argued that digitization would inaugurate a new era of "sovereign" individuals who could liberate themselves from the state.[51] Indeed, Thiel saw PayPal as taking a step in that direction. He described it as an attempt to create "a new world currency, free from all government control and dilution—the end of monetary sovereignty."[52]

In this account, the technokings of Silicon Valley would

become real kings. Private power would replace public power. This was not Musk's aspiration. His early insight into the importance of Uncle Sam and his experiences in Silicon Valley of the 1990s deepened his sense that governments were not something you should try to escape altogether. In fact, they might provide the conditions for profit-making. The magic trick of the founder was to stand on a stage built by the state and pull the future out of a hat.

On November 22, 1998, *CBS Sunday Morning* aired a report about how internet entrepreneurs were competing against Microsoft, the decade's largest and most dominant tech firm. The segment included an interview with the twenty-seven-year-old Musk. When the journalist asked him how he saw the future of the internet, he replied: "I think the internet is the superset of all media. It is the be-all and end-all of media. One will see print, broadcast, arguably radio, essentially all media folding into the internet."[53]

The metaphor was telling. In software, a superset refers to a programming language that contains the features of an earlier language but adds new capabilities. Musk believed the internet would do the same for media. It would assimilate older forms while introducing a new capability: interactivity. In the CBS interview, Musk described the internet as "the first two-way communication medium that is intelligent." Unlike a magazine, a radio, or a television set, the internet was something you could talk back to.

Years later, Marc Andreessen would famously announce that "software is eating the world."[54] Musk's internet superset was an early version of the same idea. It was also a theory of power masquerading as a description of a technological process. If everything became digitized, then authority would rest with those who controlled the code. Whoever owned the superset would have dominion over its subsets.

Thiel believed that technology was creating the conditions for escaping the state. By contrast, Musk's concept of the internet superset was a colonizing metaphor, not a secessionist one. It evoked a master system, not a parallel realm. The superset was built on hierarchy, "folding" in everything below it, in Musk's phrasing. Superset was another word for hegemony. All legacy institutions were fair prey. Muskism would carry this lesson into the next chapter of its evolution, arm in arm with the giants of flesh and steel.

3

Sovereignty as a Service

On September 10, 2001, Donald Rumsfeld held a town hall at the Pentagon. The Defense Secretary assembled hundreds of senior staffers in a large auditorium and told them he had identified an "adversary" that posed a "serious threat" to the nation. The Soviet Union was gone, but another formidable enemy remained. The enemy was the "Pentagon bureaucracy," Rumsfeld explained. It shared certain features with the vanished Soviet state: a fondness for central planning, an addiction to waste and inefficiency. These qualities made the United States woefully unprepared to defend itself. The solution was to modernize the military by embracing the "technology revolution" that was transforming the private sector: "successful modern businesses are leaner and less hierarchical than ever before. They reward innovation and they share information. They have to be nimble in the face of rapid change or they die."[1]

This was the spirit of Silicon Valley. The dot-com bubble had popped, but the internet lived on. Millions of Americans were getting online each year. Google had just turned its first profit thanks to AdWords, its new advertising service. Meanwhile, corporate America tried to keep pace with the new information age by experimenting with more flexible and networked styles of organization. Now Rumsfeld wanted to bring this mindset to the Pentagon. As he later explained, the goal was to

SOVEREIGNTY AS A SERVICE

get members of the military "to behave less like bureaucrats and more like venture capitalists."[2]

At 9:37 a.m. the next morning, a plane hit the west wall of the Pentagon, the very building where Rumsfeld had spoken. In the months and years after September 11, the Bush administration would transform the national security state in the name of fighting terrorism, with Rumsfeld in a leading role. He seized the opportunity to make the military he wanted. The "Rumsfeld Doctrine" prioritized the use of small, fast-moving ground forces supported by air power and advanced technology. To further this vision, he would outsource as many functions as possible to the private sector. Destroying the Pentagon bureaucracy would require not only an infusion of entrepreneurial thinking but the services of many more actual entrepreneurs.[3]

On October 29, 2001, Rumsfeld created the Office of Force Transformation, a small think tank tasked with promoting his agenda throughout the Defense Department. For its director, Rumsfeld chose Arthur Cebrowski, a retired vice admiral who had recently led the Naval War College. During the 1990s, Cebrowski had drawn inspiration from Silicon Valley and the work of *Wired* editor Kevin Kelly to promote what he called "network-centric warfare."[4] It was a vision of combat rewired for the dot-com era: the military would resemble the internet, with its ships and planes and tanks continuously exchanging information with one another in order to detect threats and coordinate a response. "This is warfare as a kind of interactive, multiple-player computer game," noted the *New York Times*'s Bill Keller in an admiring profile of Cebrowski.[5]

Some of the most important nodes in this new war-fighting network would be in space. The U.S. military had used satellites to spy on enemy armies since 1960.[6] But satellites had become even more valuable for warfare since the advent of GPS, which gave the Pentagon a way to not only geolocate its own forces across the globe but also to ensure that munitions reached their target. The United States first used satellite-guided ordnance in combat during the Gulf War in 1991. By the time of the U.S.-led invasion of Iraq in 2003, 22.4 percent of bombs and missiles fired at enemy targets relied on GPS.[7] The military's use of satellite-based communications also grew enormously during this period.[8]

Yet even as space became integral to combat, it remained stuck in a Cold War paradigm. Traditional geostationary satellites were like mainframes: large, expensive, difficult to program. The wars of the future required the equivalent of PCs. In May 2003, one month after the Americans entered Baghdad, the Office of Force Transformation launched the TacSat program in partnership with the Naval Research Laboratory to help meet this need.[9] The goal was to demonstrate the feasibility of small, affordable satellites that could be quickly and cheaply deployed. Cebrowski wanted to make it possible for a battlefield commander to be able to request images of a specific location and obtain the results in minutes.[10]

To achieve this breakthrough, Cebrowski would turn to the private sector. By stimulating the commercial space market, he hoped to bring down the costs of launching satellites into orbit—which was a pricey proposition so long as rockets were big, billion-dollar contraptions built by traditional aerospace firms like Lockheed.[11] The contract for carrying the first TacSat satellite into space would go to a startup that had only been around for a year. Its name was SpaceX. Its founder was Elon Musk.[12]

Chasing Uncle Sam

In 2002, Musk moved from Palo Alto to Los Angeles. "At this point, I would say I'm a little tired of the internet," he had told

CNN the previous year.[13] Back in Silicon Valley, companies were crawling out of the wreckage of the dot-com crash to finally figure out a business model. It involved building new systems called platforms that harvested user data, largely for the purpose of selling ads.

Musk had something else in mind. "I'm going to colonize Mars," he told a friend shortly before his move.[14] Space Exploration Technologies (SpaceX), the firm he incorporated in March 2002, would be the vehicle for this ambition. In October, eBay purchased PayPal. Musk, still a major shareholder, came away with $180 million after taxes.[15] It was the windfall he needed to make SpaceX a reality. At a time when Silicon Valley was entering its platform era, Musk moved in the opposite direction: doubling down on the hardest of hardware by putting millions into a rocket company.

Why did he do it? Most commentary on Musk points to his lifelong fascination with space, in particular his oft-stated desire to make humanity "multiplanetary."[16] From the start, this desire was haunted by visions of the apocalypse: one reason to colonize Mars was to ensure the survival of human civilization if a catastrophe decimated Earth, he believed.[17] Computer systems have a concept called "failover": if one component fails, another switches on. Mars would provide a failover mechanism for the kind of civilizational implosion Musk had come to expect from science fiction. (*The Hitchhiker's Guide to the Galaxy* opens with the destruction of Earth, while the story of Asimov's *Foundation* series is set in motion by the collapse of a galactic empire.)

But there is another way to understand Musk's turn to rocketry in the early 2000s. By creating the internet, the government had helped make him a millionaire. Space represented the next big publicly funded business opportunity. Speaking at

Stanford the year after starting SpaceX, he stated this rationale quite explicitly:

> I think there's perhaps an analogy here. Where just as DARPA served as the initial impetus for the internet and underwrote a lot of the costs of developing the internet in the beginning, it may be the case that NASA has essentially done the same thing by spending the money to develop some of the fundamental technologies in the beginning. And then once we can bring [the] commercial free enterprise sector into it, then we can see the dramatic acceleration that we saw in the internet.[18]

DARPA and NASA were both founded in 1958, in response to the Soviet launch of Sputnik, which caught the American establishment by surprise and spurred a significant increase in federal funding for science and technology. DARPA focused on computers and networking, making the investments that led to the internet. NASA focused on the Space Race, culminating in the Apollo missions to the moon. After the end of the Cold War, however, the geopolitical rationale for Big Science receded. In 1995, the internet was privatized. Musk intuited that space might follow a similar trajectory.

This was the Muskist logic of state symbiosis, searching for its next frontier. In space, the state would finance the development of "fundamental technologies," as Musk said. Publicly funded innovations in rocketry would be central to SpaceX's success. But that wasn't all: the state would also become a customer. "I view all of the government agencies with an interest in space as customers," he told CNN in 2004.[19]

Back in the 1950s, Silicon Valley had started out the same way: as a collection of government contractors. It even had a direct link to the space business: Lockheed's facility in Sunnyvale had been a major center for missile, satellite, and spacecraft development. (As late as 1995, Lockheed still employed 10,400 people in Sunnyvale.[20]) Musk's decision to start a rocket company struck many of his Silicon Valley contemporaries as strange, but, in some respects, it marked a return to the industry's foundations.

The move was also well timed. Musk had started his career as an internet entrepreneur in 1995, the first year of the dot-com boom. With SpaceX, he was getting in on the ground floor of another bonanza. He started the company just as the War on Terror and its associated conflicts inaugurated a new golden age of defense contracting. After 2001, military spending rose sharply and Musk put himself in a position to capitalize.[21] He started by targeting the market for launching small satellites, which would supply the initial "revenue base" from which he could gradually expand into more ambitious ventures.[22]

This plan aligned perfectly with Cebrowski and Rumsfeld's vision. Network-centric warfare required clusters of small, rapidly deployable satellites to support agile operations, while the Rumsfeld Doctrine aimed to invigorate military innovation by giving more business to entrepreneurs like Musk. His former PayPal colleague Peter Thiel followed a similar path, co-founding the data analytics company Palantir in 2003, just a year after SpaceX. (Thiel's libertarianism didn't prevent him from becoming a government contractor.) Palantir promised to discover terrorists with data-mining software inspired by PayPal's fraud detection techniques; its main investor was In-Q-Tel, a venture capital firm established by the CIA during the dot-com era.[23] Thiel would scour the world's networks for traces of America's enemies. Musk would help the Pentagon expand its orbital network to hunt those enemies from space.

On one level, this looked like a familiar story: as during the Cold War, the government was partnering with the private sector to achieve its strategic goals. But the shape of those partnerships had changed. During the Cold War, contractors played a subordinate role. They built planes, missiles, and other necessities according to government specifications and under government supervision. With the onset of the War on Terror, Rumsfeld engineered a paradigm shift. He believed that making the military more efficient and more entrepreneurial required increasing the number of contractors and expanding their range of responsibilities.[24]

In Afghanistan and Iraq, contractors often made up more than half of the total U.S. military presence—a vast increase from previous conflicts.[25] They were also involved in all aspects of the war effort, including combat, and typically operated with limited oversight: the private military company Blackwater became notorious in this regard after its mercenaries massacred seventeen civilians in Baghdad in 2007.[26] Contracting grew in the intelligence services as well. Edward Snowden was not an employee of the National Security Agency (NSA), after all, but of Booz Allen Hamilton, an NSA contractor.[27]

The expansion of the U.S. security state after 9/11, then, occurred alongside its increasing privatization. This development didn't come out of nowhere. While Rumsfeld led the charge, he was drawing on a dogma that had guided policymakers since the 1970s, namely the notion that the public sector is inherently inefficient and should outsource its functions wherever possible. This ideology, sometimes known as neoliberalism, entailed a form of privatization in which not only money but *power* flowed from public to private hands. Contractors like Blackwater took on responsibilities previously limited to sovereign entities—and were paid to do so.

SpaceX would do the same. The company got its start as a military contractor during the War on Terror. After winning the TacSat contract from Cebrowski's Office of Force Transformation, Musk managed to secure an $8 million contract from DARPA and the Air Force in 2004 to build a rocket for launching a new hypersonic weapon that could strike anywhere on Earth in under two hours.[28] The following year, in 2005, SpaceX was awarded an Air Force contract for $100 million to provide low-cost orbital launch vehicles and services.[29]

But SpaceX, like Blackwater, would also benefit from the privatization of power. The company helped pioneer a new style of aerospace contracting in which design and production decisions were no longer dictated by the state, as in the traditional Cold War model. Instead, entrepreneurs would be the ones in charge.

Agile aerospace

SpaceX signaled an evolution in the development of Muskism. The War on Terror helped nurture a new type of state symbiosis, one in which military contractors began to perform roles formerly reserved for the government. Muskism absorbed the idea that companies, by entering into such partnerships with the state, can become state-like. The goal was not to eliminate the government but to vassalize it, such that it can only exercise its authority by purchasing services from a monopoly provider. By 2025, SpaceX accounted for 95 percent of all orbital launches in the United States and more than half of all launches globally—a position that made the Pentagon, NASA, and other government agencies deeply reliant on Musk.[30] SpaceX became the de facto gatekeeper for government access to low Earth orbit. This was sovereignty-as-a-service: the logic of the modern internet platform, scaled up to the level of the nation state.

But first Musk had to actually launch a rocket. In 2002, he acquired a warehouse in El Segundo, California.[31] The building was in bad shape when he first arrived. "The ceiling boards were all hanging out, and there was water damage and God knows what," his father Errol remembered.[32] But the location was ideal: close to the Los Angeles airport, right in the heart of the country's leading aerospace hub.

The southern part of Los Angeles County had been a major producer of airplanes, missiles, and rockets since World War Two. El Segundo itself had boasted a big wartime plant that manufactured dive bombers. But by the time Musk arrived in the early 2000s, the sector had been shrinking for years. The end of the Cold War inflicted a devastating blow, as military spending fell. In 1990, there were 130,100 aerospace jobs in Los Angeles County; by 2002, when Musk came to El Segundo to start SpaceX, there were only 45,200.[33]

If the 1990s gutted Southern California aerospace, it elevated Northern California tech to a position of unprecedented prosperity. Musk came to El Segundo not only as an outsider—a dot-com millionaire who dreamed of Mars—but as an outsider who believed he knew better than the insiders. He promised to bring the magic of Silicon Valley to the making of rockets. SpaceX is "structured like a Silicon Valley technology company because that's how I know how to structure companies," he later told an audience at the National Space Society's annual conference. "Worked well the first two times, so hopefully the third time as well."[34]

This meant several things. SpaceX's headquarters would have the open office plan of a tech startup. A flat corporate structure kept management to a minimum: Musk often spoke of maintaining a high "signal-to-noise" ratio, where engineers are the signal and managers are the noise.[35] Employees were expected to work long hours at high intensity; in return, they would receive stock options and the opportunity to serve the grand mission of getting humanity to Mars. This was not what working at Lockheed or Northrop Grumman looked like. "It was this very entrepreneurial, Silicon Valley way of thinking that none of the aerospace engineers in Los Angeles were dialed into," remembered one early SpaceX employee.[36]

Most distinctive, however, was Musk's approach to production. Traditional aerospace moved slowly and carefully. Musk, by contrast, demanded a fast, iterative approach. "Build quick and learn quickly was Elon's philosophy," according to a few aerospace executives who interacted with him during SpaceX's early days.[37] Failure wasn't something to be avoided; it was how you learned. "If we're not blowing up engines, we're not trying hard enough," Musk later told a group of cadets at the U.S. Air Force Academy.[38]

This attitude was another Silicon Valley import, one that dated from the dot-com years. In the 1990s, the growth of the internet had helped foster a new approach to writing code. In the conventional model of software development, sometimes known as "Waterfall," management dictated a plan that progressed through a series of stages, from design to execution. The 1990s saw the rise of a different style, which came to be known as "Agile." Programmers took the power back from managers and rejected top-down planning in favor of rapid prototyping and constant iteration. Agile was an exuberantly anti-bureaucratic form of programming, and it proved to be an excellent fit for both the ethos and the technology of the dot-com moment.

Traditionally, companies released software on physical media. You received a set of floppy disks from Microsoft and loaded them into your computer to install Windows. With the internet, however, code could be delivered to customers anytime, even without their noticing. People didn't have to do anything to see the new version of a website—it just worked. As continuous delivery became the norm, Agile gave programmers a way to go faster.[39]

Musk took Agile and applied it to aerospace. Air Force Brigadier General Pete Worden was struck by the spectacle of SpaceX engineers at work: it looked like how "a bunch of kids in Silicon Valley would do software," he remarked. "They would stay up all night and try this and try that."[40] The company was organized to facilitate this culture. Minimizing the number of managers meant engineers were empowered to make decisions on the fly. It also reduced the distance between Musk and his employees. He liked not having to dilute his dominance through an elaborate chain of command: he preferred to rule personally. As the owner and chief executive, he was in charge. If he delegated a degree of authority to his engineers, he reserved absolute power for himself. Never content to be confined to matters of high-level strategy, he frequently drilled down to dictate the minutiae of how things were made. But whether the choice was large or small, Musk always chose quickly. As Peter Thiel would say, monarchies can move faster than bureaucracies. Especially monarchies organized along Agile lines.

Delete and integrate

SpaceX's first product would be a rocket called the Falcon 1. Its purpose was to achieve a dramatic reduction in launch costs.

In the early 2000s, putting a 550-pound payload into orbit cost around $30 million; Musk wanted the Falcon 1 to carry more than twice that—1400 pounds—for $7 million.[41] This would serve his short-term goal of cornering the launch market for small satellites. It would also serve his longer-term goal of making humanity a multiplanetary species, he said. The colonization of Mars couldn't happen until sending things into space became much more affordable. The Falcon 1 would be the Toyota Corolla or Honda Civic of rockets, he declared: cheap and reliable.[42]

This comparison held an important clue to Musk's industrial philosophy as it took shape at SpaceX. The ability of Japanese companies to produce cheap and reliable cars was in large part due to the manufacturing practices they developed from the 1950s to the 1970s under the aegis of the Toyota Production System. This system came to be known to American audiences as "lean production," a term popularized by the bestselling 1990 book about Toyota, *The Machine That Changed the World*.[43] Its influence on American capitalism was profound but uneven. Lean production was less a unified ideology than a collection of principles, and these principles would be assimilated by different sectors in different ways.

One example was software. Japanese companies emphasized continuous improvement through incremental changes and self-organized teams that made decisions for themselves. These ideas helped inspire the Agile movement in the 1990s; one flavor of Agile is even called "lean software development."[44] When it came to manufacturing, however, the American interpretation of the Toyota Production System served to accelerate existing trends toward outsourcing and offshoring. An older industrial model, associated with Henry Ford, tended to centralize production within a single firm. The Japanese took a more distributed approach, assembling the final product out of parts sourced from a set of independent suppliers. Their example contributed to the decentralization of American manufacturing, as the big integrated Fordist factory gave way to flexible production networks organized through global supply chains.

Musk's relationship to these developments was complex. On the one hand, he adopted several aspects of lean production. These included, above all, the fail-fast experimentalism of the Agile style that he encountered in Silicon Valley during the 1990s. He also shared the Japanese obsession with simplicity and elimination of waste. His instinct was always to reduce the complexity of an industrial product or process: "delete" became one of his favorite words. He encouraged his engineers to rip out a part wherever possible; if its absence caused a problem, he counseled, they could always add it again later.[45]

Yet in other respects Musk departed from lean doctrine. In particular, he came to embrace something that had become anathema to American industry by the early 2000s: vertical integration. Like other manufacturers, traditional aerospace companies made rockets from components furnished by subcontractors. Musk started out trying to work the same way. But he became frustrated with the high prices quoted by aerospace suppliers. So, he decided to produce as much as possible in-house. "If we didn't make stuff ourselves, we would be beholden to those legacy costs," he later explained.[46]

Vertical integration enabled Musk to exercise greater control over how his rockets were built, down to the last detail. "We've generally found it's more efficient to be vertically integrated than outsourced," he told an audience in 2005.[47] This efficiency was achieved by applying his relentless drive toward simplification and optimization to the entire supply chain. "Question every requirement," he told his employees.[48] In his view, the only requirements that mattered were those dictated by the laws of physics. Just because rockets had been made a certain way in the past wasn't a good reason. Just because the Pentagon and NASA wanted rockets made a certain way wasn't a good reason either.[49]

The space industry had a "very complicated regulatory structure," he complained, partly due to the long list of rules that the government required its contractors to follow.[50] One result was that rockets resembled bespoke suits: constructed by specialists according to painstaking specifications, mostly by hand. Musk, by contrast, wanted to crank out rockets on an assembly line.[51] His insistence on mass production, together with vertical integration, evoked the Detroit of Henry Ford—a comparison he himself acknowledged, albeit with caveats. "He actually mined the ore," Musk told one interviewer. "We're not that crazy."[52]

Here too he added elements that echoed the innovations of the Japanese. The SpaceX factory was laid out so that the people who designed, engineered, and built the rockets were all clustered together. "The people on the assembly line should be able to immediately collar a designer or engineer and say, 'Why the fuck did you make it this way?'" Musk explained. "If your hand is on a stove and it gets hot, you pull it right away, but if it's someone else's hand on the stove, it will take you longer to do something."[53] Short response loops and cross-functional collaboration were two defining features of the Toyota Production System.

In the early 2000s, as Musk moved from Northern to Southern California and transitioned from bits to atoms, he came up with a methodology for making stuff that would define his career as the twenty-first century's most famous industrialist, first at SpaceX and then at Tesla. It drew from different styles and eras of industrial production to arrive at a distinct synthesis that could be called "lean Fordism." He centralized in order to simplify. He integrated in order to accelerate. He fashioned a manufacturing model fit for an Agile king.

Musk had set out to build a rocket. But more consequential than what he built was how he built it. His biographer Walter Isaacson would describe a lesson Musk learned from his close friend the investor Antonio Gracias: "It's not the product that leads to success. It's the ability to make the product efficiently. It's about building the machine that builds the machine. In other words, how do you design the factory?"[54]

Musk's drive to reduce his reliance on external suppliers and to concentrate production as much as possible within the walls of the firm cut against the globalizing currents of the 2000s, which positioned the factory as a node within an international production network woven together through supply chains. Muskism, by contrast, envisions the factory as an enclave. Out of step with the 2000s, this would come in handy in the 2010s and 2020s, as SpaceX and Tesla navigated the tariffs, geopolitical tensions, and supply chain shocks of a deglobalizing world.

In and against the state

For the Bush administration, the War on Terror represented a civilizational struggle against an existential threat. Winning this struggle would, according to the Rumsfeld Doctrine, require deeper integration between the public and private sectors. Such integration would be initiated from within the state itself. Contractors became more numerous and more powerful during the 2000s because highly placed individuals within the U.S. government believed that handing more responsibility to the private sector was in the public interest. These included not just

Rumsfeld but figures like Pete Worden, who, as director of the Space and Missile Systems Center in Los Angeles, was the man who persuaded DARPA and the Air Force to take a chance on SpaceX by giving it a contract in 2004.[55]

Musk's efforts to find cheaper ways to deploy small satellites appealed to Worden because Worden had been trying to do the same thing for decades. In particular, he had worked on Ronald Reagan's Strategic Defense Initiative (SDI) in the 1980s, a program to develop a space-based missile defense program that could protect the United States from nuclear attack. One proposal, "Brilliant Pebbles," would have mobilized thousands of small satellites with heat-seeking missiles to intercept ICBMs.[56]

No SDI technology was ever deployed, and the initiative—derisively labeled "Star Wars" by its critics—came to an end in 1993. But it established the idea of space as an important warfighting domain and, more concretely, spurred the Pentagon into funding efforts to reduce the cost of putting military assets into orbit. Toward that end, in the early 1990s, Worden helped finance the prototype of a reusable launch vehicle called the Delta Clipper.[57] (Most rockets, then and now, are expendable; their components are either destroyed on reentry or abandoned in space.) Later in life, Worden came to see SDI as "the start of a new space movement" that turned away from "the traditional aerospace companies."[58] While the big SDI contracts went to the same defense "primes" that had long dominated missile production such as Lockheed and Boeing, the Pentagon also directed money to a handful of smaller firms and bankrolled experiments like the Delta Clipper.[59]

Worden was not the only SDI veteran with a connection to Musk. Others included James Maser, who briefly served as SpaceX's president, and Michael Griffin, an early adviser.[60]

Musk had even asked Griffin to come onboard as the company's chief engineer. Griffin declined, choosing to become the president of In-Q-Tel instead.[61] When Musk had first contemplated doing something in space, he hoped to "reignite the dream of Apollo."[62] But it was the dream of Star Wars, not Apollo, that loomed largest in SpaceX's first years.

To figures like Worden, SDI offered a vision where the privatization of space would help drive its militarization. The defense sector would be updated for the leaner, faster, less regulated capitalism of the Reagan years, unleashing entrepreneurial energies in the service of American global supremacy. The War on Terror revived these hopes: Rumsfeld himself had been a Star Wars devotee, as well as a relentless exponent of free enterprise.[63] SDI prefigured the kind of state symbiosis that came to fruition during the 2000s, as a reinvigorated American war machine pulled Musk into its orbit.

Musk was not simply a passive beneficiary of moves made by Worden, Rumsfeld, and other reformers, however. He also played an active role in shaping the new era of public-private partnership to his advantage. Nowhere was this clearer than in the series of events that followed from his decision to contest the awarding of a NASA contract to a competitor only two years after he started SpaceX.

In February 2004, NASA purchased flight data from a company called Kistler Aerospace for $227 million.[64] Kistler was, like SpaceX, a small private space startup, but its CEO was a NASA luminary named George Mueller, who had led the Apollo program during its heyday. To Musk, the no-bid deal looked crooked. He filed a legal complaint with the Government Accountability Office (GAO), arguing that the contract should have been awarded on a competitive basis.[65] When a Senate committee invited him to Capitol Hill to testify about the future of space exploration, he used his time to rail against the Kistler contract and demand "the opportunity to compete on a level playing field to best serve the American taxpayer."[66]

It was a risky move. SpaceX wanted to become a major government contractor. Antagonizing the government might not be the best strategy. But the gambit paid off. The GAO sided with SpaceX and NASA rescinded the contract. More importantly, the episode pushed NASA to transform its relationship with the private sector in ways that directly benefited Musk.

At the time, NASA had a problem. In January 2004, President George W. Bush announced the impending retirement of the Space Shuttle. The program had been floundering for a while; the loss of the *Columbia* the previous year, which killed all seven astronauts on board, had been the final straw. But without the Space Shuttle, NASA would no longer have the ability to reach the International Space Station (ISS). It needed alternatives.

The approach championed by Michael Griffin, the SDI veteran and former Musk adviser who became NASA administrator in April 2005, was to unleash the free market. "We believe that when we engage the engine of competition, these services will be provided in a more cost-effective fashion than when the government has to do it," he announced.[67] In partial response to the GAO incident, Griffin led the creation of Commercial Orbital Transportation Services (COTS), an initiative to develop private-sector capabilities for bringing cargo and crew to the ISS. It would proceed through an open competitive process: NASA issued an announcement and invited companies to bid.[68]

In August 2006, SpaceX won a COTS contract worth $278 million. The money would support the construction of a new, bigger rocket called the Falcon 9, along with a cargo spacecraft—the Dragon capsule—capable of resupplying the ISS.[69] The infusion of funds couldn't have come at a better time. So far, the company had brought in less than $50 million in revenue and had spent nearly all of it.[70] Musk was running out of his own cash to keep the lights on. Rumsfeld's Pentagon had furnished some early sustenance, but the Falcon 1 rocket still hadn't flown successfully.

The COTS money was a lifeline. "Absent that, you could debate whether SpaceX would have survived or not," remarked James Maser, the company's president at the time.[71] The design of the program also played to SpaceX's strengths. COTS used a "fixed-price" model, which meant a contractor is paid a predetermined sum in exchange for reaching certain milestones. This was a departure from the "cost-plus" arrangement that had dominated aerospace since World War Two, in which the government covers a company's costs plus a guaranteed profit. Cost-plus was lucrative for big firms like Lockheed, but it sapped their motivation to figure out how to make things more cheaply. Fixed-price contracting, on the other hand, incentivized the sort of relentless cost reduction that Musk implemented in his factories.[72]

Another important distinction between the two paradigms concerned the question of power. In cost-plus, the government told a contractor exactly what to do. It paid a premium in return for control. In fixed-price, however, the government provided a high-level goal but gave the firm the freedom to determine how to meet it. Relatedly, COTS contracts came with a much lighter regulatory burden. Because they were awarded under a special NASA authority that bypassed the traditional procurement process, they didn't require contractors to comply with the detailed rules specified in the Federal Acquisition Regulation.[73] This also worked in Musk's favor. Regulation was expensive: by reducing it, he could make rockets more cheaply.

SOVEREIGNTY AS A SERVICE

A company with the qualities of a country

Musk's agitation around the Kistler contract in 2004 had helped spur the creation of COTS. And COTS, in turn, marked the moment when NASA began to embrace fixed-price contracting—a development that boosted SpaceX at the expense of traditional contractors. There was a lesson here. Musk wanted to win government business by offering cheaper access to space. The optimizations he pursued on the shop floor were geared toward that goal. But that wouldn't be enough. To change the cost structure of the launch industry, he also had to change its regulatory structure.

"What can be very frustrating is that regulation is often irrational," Musk told an audience at Stanford in 2003. "It doesn't make any sense."[74] He arrived at the following solution: he would be the one to decide what made sense. And he would not be shy about exercising this authority, even if it meant challenging the law. "If the rules are such that you can't make progress, then you have to fight the rules," he said.[75] SpaceX would fight the rules constantly, whether those set by NASA, the Pentagon, or the Federal Aviation Authority (FAA).

These moves weren't just about getting the government out of the way but, more precisely, taking on its powers for himself. The same logic of privatization that enabled Blackwater to operate freely in Iraq was vesting him with powers previously unimaginable for private entities. SpaceX would enjoy the advantages of being a government contractor with little in the way of government supervision. Significantly, Musk secured this position with the support of key actors within the state: he applied pressure from the outside while sympathetic officials did what they could to accelerate outsourcing and deregulation from the inside.

These officials often framed their efforts as an attempt to increase competition in government contracting, so that markets long controlled by the defense primes could be opened up to smaller, less conventional firms. The irony is that the endpoint of this process—which began under the Bush administration but continued through the Obama administration and beyond—would be the formation of a new monopoly. SpaceX would come to dominate the launch market so completely that it became far more powerful than a firm like Lockheed ever was.

Admittedly, the company conquered the market in large part by achieving a dramatic reduction in launch costs. In this sense, reformers like Worden and Griffin were entirely correct that changing the structure of government contracting would incentivize the innovation needed to make getting to space cheaper. But this cheapness came with a cost: the state would eventually cede its sovereignty to such a degree that it was forced to buy it back in increments from a corporation.

In September 2008, the Falcon 1 finally reached orbit for the first time, becoming the first privately developed liquid-fuel rocket to do so. "There are only a handful of countries on Earth that have done this," Musk announced afterwards. "It's normally a country thing, not a company thing."[76] But Muskism is not about replacing countries with companies—it is about fusing the two. By the 2020s, Musk's dominance of the launch market meant that states needed his infrastructure to achieve their goals.

Storming heaven

With SpaceX, Muskism's pursuit of state symbiosis converged with the privatizing impulses of the early twenty-first century to find a deeper channel. The company became a global platform for national projects, a purveyor of sovereignty-as-a-service not only to government agencies within the United States but to those throughout the world. SpaceX had always sought business with other countries: as early as 2003, the company secured a $6 million deal to launch a communications satellite for the Malaysian government.[77] But its full potential as a global platform for national projects would be fulfilled by Starlink, the satellite internet service that Musk first announced in 2015.

Satellite internet had existed for decades but was plagued by slow speeds. Starlink's innovation was to bring the satellites closer to Earth—much closer. Traditional geostationary satellites orbit 20,000 miles up. To increase network bandwidth, Musk decided to place his satellites in low Earth orbit, around 400 miles up. But there was a catch: you needed a lot more satellites.

In May 2019, SpaceX's Falcon 9 rockets began launching Starlink satellites into orbit. By 2025, there were more than 8,000 of them, which accounted for two thirds of all active satellites.[78] Today, if you look up at the night sky and you see a small light moving, it's more than likely that it belongs to Musk. In less than a decade, he has transformed the heavens. Without most of us realizing it, the lower atmosphere has become a beehive of solar-winged drones speaking to one another by lasers. If Apollo symbolized the first Space Age, the second Space Age belongs to the low-flying, high-speed satellites that make up Starlink's "megaconstellations."

Starlink began its commercial service in 2021, initially focusing on areas of the United States where traditional broadband infrastructure was limited or nonexistent. But it soon began negotiating with governments around the world to secure regulatory approval and, by 2025, was available in more than

100 countries.⁷⁹ Starlink also became a government contractor, as agencies in the United States and other countries purchased satellite internet subscriptions. Another source of public money sought by Starlink has been subsidies for connecting underserved communities: in 2020, Trump's FCC tentatively awarded the company nearly $900 million to help get rural households and businesses online. While Biden's FCC revoked the award, the second Trump administration has rewritten the rules for a different federal grant program to open the spigot for Starlink.⁸⁰

Most important, however, is how indispensable Starlink has become to modern militaries. The turning point came after the Russian invasion of Ukraine in 2022, when the portable Starlink receivers became an essential means of battlefield communication and coordination by Ukrainian forces. This served as a proof of concept for further investment by the U.S. government, which had already closed a $1.8 billion contract with SpaceX in 2021 to build a military version of Starlink called Starshield that provides additional capabilities such as encrypted communications, radio and optical sensing for reconnaissance, infrared sensors for early missile detection, and the ability to locate and track objects on Earth.⁸¹ Starshield is the fulfillment of the old Star Wars dream: a swarm of cheap, nimble satellites that can ensure American space supremacy. If some are taken out, the mesh still works. This is the architecture of a distributed network—what defense officials, in an update of the network-centric warfare concept of the 1990s, have begun to call "mosaic warfare."⁸²

If the invasion of Ukraine made Starlink a household name, however, it also clarified the risks posed by SpaceX's growing power. According to his biographer Walter Isaacson, Musk disabled Starlink access within a hundred kilometers of the coast of Crimea in September 2022 in order to prevent an attack by

SOVEREIGNTY AS A SERVICE

Ukrainian drone submarines on the Russian Navy. Because the drones required internet connectivity, they didn't work without Starlink.[83] Isaacson reports that Musk was guided by a conversation he had held a few weeks earlier with the Russian ambassador to the United States, who warned him that an incursion into Crimea could lead to nuclear war. The journalist Ronan Farrow, who helped publicize the incident, observed that "there is little precedent for a civilian's becoming the arbiter of a war between nations in such a granular way."[84]

Musk later denied the story, tweeting that Starlink had already been disabled in the region so "SpaceX did not deactivate anything."[85] Isaacson followed up with a correction that wholeheartedly accepted Musk's version of events.[86] But, in a separate incident reported by Reuters, it was revealed that Musk did cut service in eastern Ukraine around the same time, crippling a planned Ukrainian counteroffensive in Kherson.[87] (Musk didn't comment on the article, while a SpaceX spokesperson dismissed the reporting as "inaccurate.") After the Polish foreign minister warned his country would be "forced to look for other suppliers" if Starlink was an "unreliable provider," Musk shot back, "Be quiet, small man."[88] When a single individual owns the infrastructure, he doesn't need to run a government to shape geopolitics.

To say that Muskism offers sovereignty as a service is not to imply it is pure exploitation. If sovereignty means not only freedom from interference but also the capacity to act, then Musk's products do empower nations. Dozens of countries have used SpaceX to put their own satellites into orbit. Spain called its SpaceX satellite launch a step toward "strategic sovereignty."[89] But this exercise of state power depends on the whims of a single man.

The wager of Muskism is that sovereignty, going forward, will be infrastructural before it is territorial—defined by access to bandwidth, compute, launch cadence, and orbital real estate as much as by borders and bureaucracies. During his dot-com days, Musk had treated money as code. At SpaceX, he treated laws and regulations the same way, as both readable and writable. If sovereignty could be purchased on a subscription basis, the terms of service would ensure that power remained in private hands.

4

Electric Autonomy

On his forty-seventh birthday, Barack Obama gave a campaign speech in Lansing, Michigan. It was August 4, 2008—three weeks before he would officially accept the Democratic nomination for president. The United States, he warned, suffered from an "addiction to foreign oil"—one of the most dangerous and urgent threats the nation had ever faced. Burning fossil fuels warmed the planet, driving drought, famine, and disaster. But climate change was only part of the problem. The world's oil supply was running out. As it did, gas prices would keep rising—oil had just hit an all-time high of $145 a barrel—squeezing consumers and businesses. Meanwhile, oil-rich Middle Eastern countries would see their power continue to grow, contributing to further "instability and terror," and drawing America deeper into entanglements with "some of the world's most unstable and hostile nations." In the summer of 2008, the wars in Afghanistan and Iraq were still raging.

The solution, in a word, was autonomy. Energy independence had been a goal of American presidents since Nixon, Obama acknowledged. But by embracing "clean, affordable, renewable energy," the United States could finally achieve this ambition, and liberate itself from climate change, rising gas prices, and the quagmire of the Middle East. It would also mean jobs. Lots of "good, union jobs" to revitalize Michigan, the historic home of the American auto industry, which would be retooled to make electric vehicles and hybrids. "For a state that has lost so many and struggled so much in recent years, this is an opportunity to rebuild and revive your economy," Obama declared.[1]

The next month, Lehman Brothers collapsed. The financial crisis that had been unfolding since 2007 was accelerating. In November, Obama won the presidency. He promised to lead the country out of the cataclysm and "reclaim the American Dream." Central to this project of national renewal was federal investment in clean energy, what the columnist Thomas Friedman was calling a "Green New Deal."[2] Among its beneficiaries would be Elon Musk.

Walled gardens

The story of Tesla has been told many times—and almost always as a story about cars. But if we take Musk at his word, Tesla is not just about cars. "It's far more than what people normally think of as a car company," he explained to his shareholders in 2024. Then he showed them a map of the Tesla ecosystem.[3] It pictured a grid. A cherry-red sedan—like the one Musk later sold to Trump on the White House lawn—was only one of thirteen components. Some were vehicle-related—insurance, self-driving—while others involved AI, robotics, solar panels, energy storage, and even the refining of lithium, a key element in modern rechargeable batteries. Arrows linked everything together, suggesting an integrated system.

What ties together the various pieces of this system? From the beginning, Tesla has sold the promise of electric autonomy: the notion that renewable energy can promote self-reliance.

This idea can be scaled up or scaled down. Scaling up to the national level, electric autonomy means using EVs to secure national energy independence, freeing countries from dependence on foreign oil and the geopolitical entanglements that come with it. Scaling down to the personal level, it means using Tesla's cars, solar panels, and energy storage systems to foster individual resilience against grid failures and natural disasters in an increasingly unstable world. Tech analysts use the term "walled garden" to describe a closed digital environment owned and controlled by a single company—Apple's App Store is the classic example.[4] The walled garden of the Tesla ecosystem is a technology for hardening the walls of the nation and the home.

Over the years, Tesla has marketed different models of electric autonomy, adapting its lineup to shifting political conditions. Along the way, it has made lucrative use of the Muskist motifs explored in previous chapters, from state symbiosis to financial fabulism to fortress futurism. If Tesla is the company that most defines Musk in the eyes of the public, it also offers the most complete example of Muskism in action. Three developments converged to make Tesla's success possible.

The first is the American experiment with green capitalism in the 2000s, as championed by the early Obama administration. The aspiration was to trigger a clean-energy boom that could pull the country out of the Great Recession while simultaneously breaking the oil addiction that had contributed to the disastrous military adventurism of the Bush years. Naturally, this would be achieved through a public–private partnership: the government would give money to entrepreneurs to build the post-fossil fuel future, creating opportunities for state symbiosis that would prove indispensable for Tesla. If SpaceX had benefited from the War on Terror, Tesla would benefit from the backlash.

The second enabling factor for Tesla was the rising dominance of Silicon Valley. In the 2000s, the tech industry began to belatedly fulfill the promise of the dot-com era by building profitable internet businesses, from Google to Amazon to Salesforce to Netflix. The tech sector's growth accelerated in the aftermath of the financial crisis, as falling interest rates from 2007 onward encouraged investors to look for higher returns by plowing money into venture capital funds and into the stocks of publicly traded tech firms. Meanwhile, the proliferation of smartphones, social media, and cloud computing in the 2010s created new revenue streams. By 2013, the CNBC personality Jim Cramer was talking about the FANG trade: Facebook, Amazon, Netflix, and Google.[5] Tech became synonymous with skyrocketing valuations and fat profit margins. Tech had become "Big Tech."

From the start, Tesla would tout itself as a tech company. Headquartered in Palo Alto, the company emerged during the "cleantech" boom that swept Silicon Valley in the 2000s. Tesla would also try to make driving more digitized, as evidenced by the decision to install large touchscreens in its cars as early as 2012. Three years later, it launched Autopilot, a limited form of lane-keeping and cruise control enabled by the crowdsourced data logged by the car's camera and sensors, and centrally processed and analyzed by the company. "Carmakers need to think of their cars as connected devices," Musk said at the time, "like your cell phone or laptop."[6]

The third ingredient behind Tesla's success was the rise of China. By the mid-2010s, China's upward trajectory had begun to generate anxiety within the American political establishment—especially as it watched China's investment expand overseas with the Belt and Road Initiative. The relationship between the two countries started to deteriorate over the course of Obama's presidency but tensions rose sharply during Trump's first term, as a trade war officially erupted in 2018. The following years would see a succession of tariffs and export bans, with Biden continuing what became a bipartisan crusade to constrain Chinese power. These moves marked a retreat from the free-trade consensus of the 1990s and the 2000s. So did the growing interest among American politicians in reshoring the country's manufacturing base through reinvestment and protectionism—a scheme that gained new urgency after the supply chain disruptions of the Covid-19 pandemic exposed the fragilities of a highly integrated world.

Deglobalization would play to Tesla's strengths. When Musk began to embrace vertical integration in the 2000s, it defied conventional wisdom. By the 2020s, it looked prescient. The siege mentality of apartheid South Africa, untimely in its own era, helped foster a strand of Muskism that harmonized perfectly with the zeitgeist. The factory enclave of fortress futurism turned out to be an industrial form well suited to a fracturing world. Tesla didn't just weather the storm; it capitalized on the chaos. Musk responded to the new reality of interstate competition by making everyone a customer. Like SpaceX, Tesla would provide a global platform for national projects. In the dream of electric autonomy, Muskism achieved its fullest expression to date.

The old green capitalism

Tesla was founded in 2003 by two engineers who wanted to wean the United States off its addiction to foreign oil. Martin Eberhard and Marc Tarpenning had made a bundle selling their e-reader company a few years earlier and used the proceeds to kickstart their next Silicon Valley startup: a company that aimed to sell an electric sports car to wealthy, eco-minded Northern Californians. In 2004, they contacted Musk to see whether he would be interested in investing. He loved the idea and came onboard as the company's primary funder and chairman, eventually taking over as CEO in 2008 after both founders departed.[7]

These were the years of the cleantech boom in Silicon Valley. While some entrepreneurs were building the first internet platforms, others were trying to make money from solar panels, batteries, and other technologies of the renewable transition. An early booster was Al Gore, a longtime tech ally who had shepherded early internet legislation through Congress and overseen its privatization as vice president in the 1990s.

Gore built a second act in Silicon Valley after losing the presidential election in 2000. He advised Google and served on the board of Apple. But he became best known as an evangelist for climate action, the subject of his 2006 Oscar-winning documentary *An Inconvenient Truth*. The following year, Gore joined the storied Silicon Valley venture firm Kleiner Perkins at the behest of John Doerr, a legendary investor who had made a fortune off the internet.[8] Doerr saw Silicon Valley's next frontier in clean energy. "Green technologies—going green—is bigger than the internet," he announced in a 2007 TED talk. "It could be the biggest economic opportunity of the twenty-first century."[9]

This opportunity attracted a significant amount of capital. In 2001, total venture capital investment in cleantech amounted to only $365 million. By 2008, it had surged to $6.65 billion.[10] Still, cleantech's leading figures understood that the private sector couldn't do it alone. "What we are going to have to put in place is a combination of the Manhattan Project, the Apollo project, and the Marshall Plan, and scale it globally," Gore told *Fortune* after joining Kleiner Perkins.[11]

The government obliged. Congress passed energy bills in

2005 and 2007 that provided various forms of support for cleantech, from energy efficiency mandates to tax incentives and loan guarantees.[12] But the onset of the financial crisis in 2007 and 2008 hit the industry hard. Musk called it "market Armageddon."[13] In this dismal moment, Obama's victory offered hope. The incoming administration was full of thinkers who saw the crisis as an opportunity—a chance to rewire the American manufacturing base toward renewables and electrification. Obama was the first "internet president," after all, whose savvy use of the web had helped propel him into office. The tech industry loved him. Now he could be the savior of Silicon Valley's next act.

Musk always had an instinct for putting his entrepreneurial energies into sectors that could benefit from the munificence of Uncle Sam. As he had told an audience at Stanford in 2003, the federal government had financed the basic technologies of both the internet and space. Zip2, X.com, and SpaceX were all creatures of the state.

Tesla would be no different. It would take advantage of an opportunity for state symbiosis that arose partly from a sense of discontent with the War on Terror. Musk's top lobbyist at Tesla was Diarmuid O'Connell, former chief of staff for George W. Bush's assistant secretary of state for political–military affairs. Each morning at the State Department, he read the official report about the soldiers who had just died in Iraq and Afghanistan. "I kept thinking, this is insane. Why are we in this place?" he later recalled. He came to believe that dependence on foreign oil was the answer.[14] Seeing the War on Terror from the inside had given him a passion for renewables.

O'Connell lobbied for legislation to give tax credits to buyers of EVs, making Teslas cheaper for consumers. It was passed into law in the final year of George W. Bush's presidency as the Energy Improvement and Extension Act of 2008. But his real coup was helping to secure a $465 million loan from the Department of Energy, announced in June 2009. It brought Tesla through the near-death experience of 2008, which Musk called "the most painful year" of his life.[15]

The money was awarded under the Advanced Technology Vehicles Manufacturing Loan Program, an initiative begun under President Bush in 2007 to help automakers and parts suppliers manufacture more fuel-efficient vehicles. Congress didn't fund the program until 2009, giving Obama an opportunity to funnel capital to firms that could fulfill his vision of green capitalism. As one of the very few American companies making EVs at the time, Tesla was an obvious choice—although it probably didn't hurt that the venture capitalist Steve Westly, a Tesla investor and board member, also happened to be a major fundraiser for Obama and co-chair of the campaign's California operation.[16]

The Tesla loan generated controversy. Some grumbled about taxpayer dollars going to a company producing a $100,000 car with fewer than 200 units in existence by the time Obama took office. The *New York Times* wondered if it should be called the "2008 Bailout of Very, Very High-Net-Worth Individuals Who Invested in Tesla Motors Act." "Only the rich can afford it," the headline asked. "Should taxpayers back it?"[17] Yet Tesla's defenders argued that starting high was a strategic choice. Selling a luxury model first was a way to shed the stigma of the EV as a glorified golf cart and pave the way for more affordable models down the line. In fact, Tesla had applied for the Department of Energy loan precisely so that it could scale up the manufacturing of its newly announced Model S, a sedan that would cost $49,900 (after tax credits).[18]

And Tesla wasn't alone in seeking federal aid. It belonged to a crowded field of struggling cleantech startups that hoped

Obama would come to their rescue. "Silicon Valley has mocked the government for decades and is now completely dependent on it," remarked one analyst. "They can't get a project off the ground without these loans."[19] More charitably, the new Obama administration was simply doing industrial policy. It was using its resources and its leverage to nudge the American economy toward a less carbon-intensive future.

The same spirit animated California's influential zero-emission vehicle (ZEV) program, created in 1990, which required automakers to sell increasing numbers of vehicles with no exhaust emissions. If they couldn't meet the mandated threshold, they had to buy credits from firms (like Tesla) that exceeded the threshold. The ZEV program, which went on to be adopted by sixteen more states and the District of Columbia, became a major moneymaker for the company, helping it become profitable in August 2009, only one year after the release of its first car, the Roadster. Over time, this line of business would grow dramatically. At its peak in 2024, about 40 percent of Tesla's net income came from selling credits—an indirect but essential subsidy.[20]

No war for oil

In retrospect, the first Obama term stands out as a rare moment of consensus on the need to reduce U.S. dependence on fossil fuels, especially those imported from the Middle East. Musk did believe that burning fossil fuels was heating the planet, though that wasn't the only reason he wanted to electrify the auto industry. Another motivation was geopolitical: the "true cost of gasoline," he said in an interview, included "the auxiliary effect of wars."[21]

By the time Obama won the presidency in 2008, this was not a fringe view. No less a figure than former CIA director James Woolsey, who had served both Republican and Democratic presidents over his long career in government, championed EVs as a way to end American reliance on foreign oil.[22] Just as drones could kill America's enemies without American boots on the ground—drone operations would rise sharply under Obama—EVs could insulate the country from the entanglements of distant desert conflicts. The insect-like hum of the electric motor was the sound of a new age of American power.

To Obama's America, Musk promised electric autonomy. Just as with SpaceX, he would disrupt a legacy industry with the mindset and methods of Silicon Valley—as well as its absence of traditional aerospace and auto industry encumbrances like labor unions. SpaceX had found a foothold by advancing the national security goals of the Bush years. Tesla would do the same in the Obama era. The satellite and the EV were two expressions of a single imperative: to secure the homeland.

In 2010, Tesla used the money from the Energy Department loan to open its first factory. It was located in Fremont, California, twenty-seven miles from the company's headquarters in Palo Alto. The building formerly housed a joint venture between General Motors and Toyota called New United Motor Manufacturing, Inc., formed in 1984 to help U.S. automakers absorb the innovations of the Toyota Production System. When the plant closed in 2010, Tesla purchased it at a significant discount.[23] As Musk prepared to mass manufacture cars for the first time, he brought the lessons he had learned from mass manufacturing rockets in Southern California. He would apply and refine his philosophy of "lean Fordism," combining Fordist mainstays like vertical integration with the agility of the Japanese industry.

The month after Tesla agreed to buy the factory, Tesla went

public. It was the first IPO by an American car company since Ford in 1956. That night, Musk threw a party at the Fremont plant. Raising a glass, he made a toast. "Fuck oil," he said.[24]

Then, in the early 2010s, cleantech collapsed. While the first Obama administration put more than $90 billion into clean-energy projects, the most in history to that point, it still fell short of what the sector had hoped for. One obstacle to further support was the Republicans, who retook the House in November 2010. The next year, a solar company called Solyndra that had received a $535 million federal loan guarantee went bankrupt. Republicans seized on the story, using it to attack Obama as his re-election campaign began. Solyndra became "a favorite whipping boy of conservatives warning against the perils of industrial policy," in the words of the scholar Joan Fitzgerald.[25] It made for an easy tombstone to mark the end of that brief, hopeful moment.

The real killer of cleantech wasn't politics, however, but technology. Beginning in the mid-2000s, the "shale revolution" transformed the U.S. energy landscape. Using techniques of "hydraulic fracturing"—fracking—that had been developed with federal R&D money, a set of smaller firms unlocked vast, previously inaccessible reserves of oil and natural gas. Between 2007 and 2016, oil production in the United States increased 75 percent and natural-gas production increased by 39 percent. In 2018, the United States became a net exporter of oil for the first time in almost seventy-five years.[26]

Energy independence had been achieved—not by greening capitalism, however, but by finding a way to greatly increase the extraction of fossil fuels. The shine faded from cleantech. Investors fled: by 2011, an MIT report concluded, "almost all of the 150 renewable energy start-ups founded in Silicon Valley over the past decade had shut down or were on their last legs."[27] What had been pitched as a revolutionary shift soon came to be seen as little more than a costly detour—a high-tech boondoggle. Yet the counterexample of Tesla always stalked this narrative. It offered living proof that industrial policy—what economist Mariana Mazzucato calls the "entrepreneurial state"—could work.[28]

Tesla repaid its loan to the Department of Energy in 2013, nine years ahead of schedule. Two years later, Musk made a symbolic move when he expanded Tesla's Fremont factory to occupy the former footprint of Solyndra's failed solar operation. More quietly, Tesla had already absorbed an old UAW union hall, a nod to the changing face of American industry.[29] From the start, Tesla was a fiercely anti-union employer.

How did Musk do it? How did Tesla ride the wave of the cleantech boom but then manage to outlive its collapse? Musk would say that everything came down to the design of the factory. The innovations of lean Fordism unlocked efficiencies that helped Tesla survive. To go further, however, Musk would need to own more of the supply chain.

New paradigm

It is rare to watch a paradigm of political economy shifting in real time. Yet there is an episode of *The Charlie Rose Show* from 2011 where this happens. Sitting at the table across from Rose was Musk, who had just turned forty. He represented the future: Tesla had gone public the previous year. Representing the old guard was Bob Lutz, a seasoned auto executive who had most recently served at General Motors. Lutz spoke for a sector that appeared to be in steep decline: the financial crisis had almost killed the American auto industry. General Motors and Chrysler, titans of postwar capitalism, had entered bankruptcy in 2009.

During the interview, Lutz dismissed the idea of man-made climate change. When asked about the government bailout, he grumbled that he had feared Chrysler would be forced to make a bunch of "greenmobiles." Musk, by contrast, was speaking a different language. Clean-shaven with a Tintin lift to his forelock, he wore the starched white shirt of a high-school debater. One of the charms he possessed in this period of his life was his ability to explain complex engineering problems with the clarity of a gifted teacher. When asked about climate change, he hesitated at first, not wanting to offend his climate-denying elder conversation partner. Then he described humanity as conducting a vast experiment. We were pumping carbon into the atmosphere at an unprecedented rate and hoping it wouldn't cook the planet, he said.

Musk was also forthright about the challenges he faced at Tesla. As he said in an earlier interview with Charlie Rose, "the single most important thing" was the battery.[30] There are many kinds of batteries. The Duracells or Energizers that power home appliances are alkaline batteries made of zinc and manganese. The chunky rectangle that jolts internal combustion engines into life is made from lead plates submerged in a bath of sulfuric acid. For a variety of reasons, neither is suitable for EVs. In the 1990s, some manufacturers began producing electric cars with nickel–metal hydride batteries. This proved a more viable option, but still had serious limitations, such as relatively low energy density and charging efficiency.

The breakthrough came with the arrival of the modern lithium-ion battery. It made its commercial debut in 1991 with

Sony's CCD-TR1 camcorder. Over the next decade, lithium-ion batteries became a staple of consumer electronics. The laptops that achieved ubiquity in the new millennium used lithium-ion batteries. So did the iPhone, first introduced in 2007. But it was Tesla that proved the value of the technology for EVs.

Lutz acknowledged this in their conversation. The common wisdom, he said, was "lithium ion isn't going to work"—but Tesla was producing a car with excellent acceleration and a 200-mile range. Musk had seen how the convergence of new technology and new consumer demand had created an inflection point in the automotive industry, still one of the world's largest manufacturing sectors.

How to make a battery

To turn the battery that powers your laptop or phone into one that powers a car involves a large leap of scale. Tesla's inaugural product, the Roadster, the first highway-legal and serial-production all-electric vehicle to use lithium-ion batteries, had 6,831 individual cells.[31] Where would they come from?

At first, Tesla sourced its batteries from external suppliers. For the Roadster, the initial process involved buying lithium-ion cells from Japan, sending them to Thailand to be assembled into battery packs, and then shipping the battery packs to Palo Alto to be put into cars.[32] This might seem unnecessarily complex, but it was standard operating procedure for American companies in the early twenty-first century. The ethos of the era was epitomized by Steve Jobs, whose iPhone famously bore the inscription "Designed by Apple in California. Assembled in China" from the very first unit in 2007. For most manufacturers, the question of where components and materials would come from was an easy one: the open market.

The 1990s and 2000s represented the peak of this free-trade consensus. In 1995, the World Trade Organization (WTO) was established with the goal of promoting trade among its members. The accession of China to the organization in 2001 led to more offshoring and outsourcing by U.S. companies, accelerating a decades-long trend. From the fall of the Berlin Wall in 1989 to 2008, total world exports grew over tenfold in nominal terms.[33] A single product might cross several borders during its manufacture, as supply chains stretched across the globe. By the late 1990s, it was not unusual for a consumer good as simple as a T-shirt to be made in multiple countries, as producers chased low wages and cheap materials.

One of Musk's peculiarities was his refusal to rely on the global market to deliver the critical inputs his products required. He preferred to make as much as possible in-house. This was the logic of fortress futurism: a model of industrial self-reliance that he first implemented at SpaceX. At Tesla, however, he would take vertical integration to new extremes. His ambitions were so large in scale that the existing supply chains might not be able to bear the pressure he planned to place on them. By 2013, Tesla's chief technology officer calculated that if the company wanted to run the Fremont factory at full capacity, the battery demand alone would match the *entire global output* of lithium-ion batteries.[34] They couldn't cross their fingers and hope the market would catch up—they would have to build the capacity themselves.

Like Henry Ford a century earlier, Musk saw the need to take control of the process from the ground up, from the extraction of raw materials to the finished product. A new kind of Fordism would have to be constructed. To make its own batteries, Tesla would have to build, as Musk put it in 2013, a "truly gargantuan factory of mind-boggling size"—the Gigafactory.[35]

If the company's early adventures were a textbook example of the entrepreneurial state in action—an embryonic form of green industrial policy—the next step, the creation of the Gigafactory, followed a more conventional script: forum-shopping and interjurisdictional competition. Musk launched a national PR campaign, dangling the prospect of a massive battery plant before multiple states in the hope of extracting tax breaks and other incentives.[36]

By 2014, the winner had emerged. At a press conference with Nevada's governor, Musk announced that Tesla would build its first Gigafactory in Sparks, just outside Reno. Nevada was close enough to Fremont for logistical convenience—about a four-hour drive—but, crucially, the state has no corporate income tax, while California's is nearly 9 percent. Labor laws are also more relaxed in Nevada, alongside its looser environmental regulations. While California has long maintained labor-friendly protections, Nevada has been a right-to-work state since 1953. Overtime rules are also more favorable to workers in California, where stricter standards govern hours and compensation.

In 2015, a Tesla supplier summarized the advantages of Musk's approach in a conversation with a journalist from the *Financial Times*. Once, he recalled, Musk spotted a production problem on a Friday and demanded his engineers work through the weekend to have it fixed by Monday. "If it was Ford, you've got labor unions," the supplier said. "This would have taken two years."[37]

Gigafactory against globalization

The first Gigafactory opened in 2016, producing lithium-ion batteries at scale. Its logic was simple: to secure Tesla's supply chain and reduce dependence on external producers. The timing coincided with the Paris Agreement, which pushed global demand for renewables upward and accelerated a scramble for the technologies and materials of the green transition. In hindsight, it was a tipping point for the world economic order, as countries long reliant on globalization began to pay more attention to supply chains and prioritize domestic control of resources. Muskism rhymed with the shift.

China was the pacesetter. After years of soaring emissions, Beijing set out to dominate renewables through joint ventures, technology transfers, and relentless competition among domestic firms. By 2017, the Chinese government announced a new mandate requiring automakers to earn credits for producing "new energy vehicles" such as all-electric and plug-in hybrid cars, as part of a broader push to develop the country's EV and battery industries. Progress was swift. A battery company called Contemporary Amperex Technology Ltd, or CATL, outpaced Tesla's Nevada output within a few years of the Gigafactory's opening.[38] China's "big green bang" was tied to a broader push for strategic self-reliance, to be achieved through industrial policy at home in the form of the Made in China 2025 program and infrastructure investment abroad through the Belt and Road Initiative.[39]

Nervous about China's rising economic and political influence, the U.S. responded with tariffs, export bans, and a rhetoric of reindustrialization. The free-trade settlement of the 1990s and 2000s was being replaced by a new bipartisan consensus on *geoeconomics*—a refusal to entertain the illusion that the world of property and the world of government could be insulated from one another.[40]

Musk was made for this moment. He had been reshoring before it was cool. His emphasis on shorter supply chains and vertically integrated production perfectly anticipated the deglobalizing currents that gathered speed in the second half of the

2010s. He never lost sight of how the power to build and the power to rule were intertwined. During the first Trump administration, he faced a strategic decision. He could double down on the U.S. market, embracing the American side of the deglobalizing world order. Alternatively, he could make the case to China that Tesla could be a partner in its national project of economic self-sufficiency and strategic autonomy from the United States. There was a lot of money at stake. China is the largest car market in the world—Tesla could not afford to be excluded from it.

Faced with a choice between the two competing superpowers, Musk chose both. In 2018, he announced that the second Gigafactory would be built in Shanghai. The next year, he secured $1.6 billion in low-interest financing from a consortium of Chinese banks.[41] The facility would produce cars, not batteries. In fact, the batteries would come from CATL. By manufacturing Teslas inside China, the company could jump the tariff walls.

By the late 2010s, the contours of Musk's geoeconomic strategy were coming into focus. He would build Chinese cars for the Chinese and American cars for the Americans. In a fracturing world, he would help both countries pursue their projects of technological self-reliance and self-strengthening. And he wouldn't stop there. As Musk told his shareholders in June 2019, his goal was "a car factory on each continent."[42] Five months later, he announced that a new Gigafactory would be built in Brandenburg, just outside the German capital of Berlin.

This gave him a foothold in the heart of the European Union, the third-largest economy in the world after the United States and China. The timing was significant. The EU declared its intent to develop a homegrown supply chain for lithium-ion batteries in 2017. The following year, CATL announced it would build its first European battery factory in Germany. Years later, another in Hungary would follow. "I really think of this issue in terms of sovereignty," French finance minister Bruno Le Maire told the *Financial Times* in reference to European battery efforts. "Mobility is a matter of sovereignty."[43] In a time of fragmentation, Muskism offered a template: sovereignty through supply chains, autonomy through electrification. By selling resilience to all sides, Musk ensured that no matter who gained the upper hand, he would win too.

Dark green future

Tesla had always traded on a vision of a better future. That was its brand, pitched to buyers with the disposable income to purchase its products. Musk himself embodied this futuristic optimism. From 2008 to 2020, during a period of cheap credit and abundant liquidity, his eccentric style charmed most observers. Tesla shareholder meetings were dotted with digressions on obscure bits of technology, but the overarching narrative was clear: with better consumer choices and enlightened policy, the world was moving toward a cleaner, more livable future. But the science never really pointed in that direction. Among those in the know, it had become increasingly clear that the accumulation of carbon in the atmosphere had set in motion changes that would be virtually impossible to reverse. By the turn of the twenty-first century, the real conversation to be had was no longer about avoiding climate change—it was about mitigation and adaptation.[44]

In other words, the future would not simply be a greener version of the past. It would require reckoning with coastal erosion, intensifying weather events, a secular rise in droughts, uncontrollable wildfires, the exhaustion of groundwater, and an increasing frequency of grid failures due to extreme heat or unexpected cold. The green future, if it arrived at all, would also be a grimmer one.

There were early signs of this reality, and Musk seemed attuned to them. In places like Western Australia, where heatwaves caused blackouts, and in Puerto Rico, whose grid was devastated by Hurricane Maria in 2017, he publicized his products as solutions to structural crises. In Australia, Musk installed what was then the world's largest lithium-ion battery to back up local grids.[45] In Puerto Rico, he sold home batteries to help people keep the lights on.[46] This wasn't the sleek Roadster or the winged Model X SUV. What Musk brought to these places was Tesla Energy.

Tesla Energy is the division of the company that specializes in home energy solutions. Its installers will put solar panels on your roof and you can even obtain financing from Tesla to pay for them. Also available is the Powerwall, Tesla's home battery system, which can store excess solar energy and provide backup power during blackouts. The Powerwall came online as a product in 2015, and it became one of the main outputs of the Nevada Gigafactory. "Tesla's quest to disrupt a trillion-dollar car industry," Musk told reporters, "offers an adjacent opportunity to disrupt a trillion-dollar electric utility industry."[47] This version of Tesla wasn't about mobility; it was about resilience. It was about protecting your home from an increasingly unstable world by turning it into a fortress powered by solar panels and batteries. In 2019, the Nevada Gigafactory started producing the Megapack, a scaled-up version of the Powerwall that can be used as a backstop to entire energy grids or to provision "microgrids" in remote areas.[48]

The turn from the blue-sky optimism of the personal vehicle

to the storm clouds of power outages tracked a shift in the zeitgeist. Supply chains buckled in 2020 under the weight of coronavirus containment measures, and Musk's mood darkened even as Tesla's stock reached unprecedented heights. When the Cybertruck finally went on sale in November 2023 after being teased for years, it felt like the material embodiment of Tesla's new ethos. The playful glitz of the cherry-red Roadster was gone. In its place stood a hulking, unpainted stainless-steel vehicle—more armored shell than sports car. It echoed the Hummer's strange journey from battlefield to boulevard in the 1990s, but with an even more overt embrace of survivalist aesthetics. This was not a vehicle for a better world. It was a vehicle for whatever came after.

Here was another variation on the theme of electric autonomy. Musk wasn't just selling sovereignty to nations; he was selling it to individuals. Standing in front of Tesla shareholders in 2021 with a bandana around his neck like a rakish Labrador, he described a snowstorm that he had recently experienced in Austin, Texas. It took out the power, so he couldn't turn on the lights or connect to the internet. But if he had installed Tesla solar panels and a Powerwall, and had subscribed to Starlink, SpaceX's satellite internet service, he would have had everything he needed. "We're seeing increased extreme weather events," Musk pointed out. Fortunately, Tesla provided "all the things you need for a prepper." "If Doomsday comes, it could be helpful."[49] The value proposition was simple: in an era of ecological and institutional breakdown, you'll be fine in your Tesla dome.

Musk's shifting rhetoric in the early 2020s seemed to reflect a loss of innocence regarding clean energy, aligning instead with a more dystopian view of the future. The humanitarian rhetoric of the Obama years faded. As climate conflict and crisis made the world a more ruthless place, not all humans would survive. Musk hoped to go one step further and eliminate humans altogether. The Cybertruck, as the name suggests, was inspired by the cyberpunk subgenre of science fiction. Cyberpunk futures are characterized not only by extreme levels of social inequality but also by the disappearance and disintegration of the human through automation and augmentation.

It was no accident, then, that the introduction of the Cybertruck coincided with a broader shift: in the early 2020s, Musk began emphasizing that Tesla was an artificial intelligence and robotics company. On 19 August 2021, at Tesla's first "AI Day," he announced Optimus, the company's humanoid robot, named for the *Transformers* character. "Tesla is arguably the world's biggest robotics company," he said, referring to his cars as "semi-sentient robots on wheels."[50] He became increasingly focused—along with the analysts who sustained his stock's meteoric rise—on the pursuit of fully self-driving vehicles, from "robo-taxis" to autonomous semi-trailer trucks, which promised to remove the need for human drivers.

Musk's focus has always been on "the machine that builds the machine." His fantasy of the future is, in a way, a fantasy of the factory. In 2016, he first shared his goal of creating what he called an "alien dreadnought"—a so-called "dark factory" with no human workers at all.[51] One can imagine an extension of his model of vertical integration where fleets of self-driving trucks haul raw materials like lithium from sites within the United States. They would be offloaded by robots into a factory where other robots would transform them into battery cells, which would then be used to power not just the self-driving vehicles but the robots themselves. The entire system would run as a closed loop.

In this vision, the factory isn't just the means of production—it becomes the world itself. The entire chain of life and labor is internalized into a single, seamless system: a sovereign factory state. This was electric autonomy in a fully automated form, but it raised a deeper question: autonomy for whom?

In the most extreme version of Muskism, the human is no longer the subject of history, but its discarded scaffolding. The product is not a car, or even a robot. The product is a self-regulating infrastructure in which the world—like the factory—runs on its own. No operators, no workers—just endless production, unfolding without interruption or intervention.

In February 2020, as the world began to shut down, the Dutch architect Rem Koolhaas and his Office for Metropolitan Architecture (OMA) opened an exhibition at the Guggenheim in New York. Part of the show was devoted to Musk's first Gigafactory in Nevada, which is located in an industrial park known as the Tahoe Reno Industrial Center (TRIC). Koolhaas had been a legendary observer of New York. His 1978 book, *Delirious New York*, was a celebration of the city's grid plan, which allowed for endless experimentation within the most elegant of constraints. Later, he applied a similar lens to China's special economic zones in a massive book, written with his students at the Harvard Graduate School of Design.[52]

In TRIC, he saw a continuation of these efforts at organizing space. Koolhaas limned it in blank verse:

Because it takes place in the countryside, TRIC is a stealth revolution . . .
    The buildings here are not for humans but for things and machines.
    Thousands of years of architectural and cultural history are ditched.

Debates, predictions, ideologies thrown overboard. It is post-human.

These structures are based strictly on codes, algorithms, technologies, engineering, and performance, not intention.

There is no formalized entrance; there are no users, only robots.

Not to hinder process is the one ambition.[53]

It became harder to ignore the irony: Tesla, once a product pitched as enabling human mobility, now seemed to prefigure a future with no humans at all.

PART TWO

Cyborg

The first half of this book began with the idea that historical change is an uneven process. The future always arrives too early, sprouting from the subsoil of the present. The foundations of Muskism built over the last four chapters revealed not only an ability to spin tales about the future but also an attention to geography and the material problems of making. As an engineering ideology, it genuflected to the laws of physics and incorporated attitudes toward the factory that reached back to Musk's native South Africa and to Henry Ford before it. The vertical integration of Muskism was out of step with prevailing trends in the early 2010s but foreshadowed the deglobalizing currents of the 2020s.

The second half of this book takes up a more recent feature of Muskism: its preoccupation with cyborgs. Back in the late 1990s, during his days as a dot-com entrepreneur, Musk had spoken of the internet as a "superset." He envisioned the network devouring the world. This is in fact what happened over the course of the 2000s and the 2010s. The rise of social media and smartphones pushed more of our lives online. Meanwhile, the extraordinary profitability of the platforms turned the tech industry into the leading sector of American capitalism.

The digitization of everything, and the new power arrangements it produced, would shape the next stage of Muskism. Starting in the mid-2010s, Musk became convinced that our growing entanglement with our devices augured a larger transformation: humans were becoming cyborgs. Biological and digital systems had begun to merge, forming what he called "cybernetic collectives" of networked intelligence. His role, he believed, was to accelerate this merger. But there was a danger. The cybernetic collective, like any networked system, could be infected. It was not enough to speed up our fusion with the machines. Musk also had to take control of the interfaces where this fusion was taking place—from social media to neural implants to artificial intelligence—to ensure that the right kind of cyborgs were being made.

"I bought this before Elon went crazy" read the bumper sticker that began to appear on Teslas in early 2025. It was a reference to the fact that Musk's public behavior had grown more bizarre since 2020 while his politics swerved to the right, culminating in his close partnership with Donald Trump. But the bumper sticker raises an interesting question. When did Musk supposedly go crazy? And is insanity the best way to understand what happened to him?

The second part of this book suggests that the answer lies in what we call Muskism's "cyborg turn." The turn did not mutate out of one man's cyberpunk delusions, but out of a bipartisan consensus about where America needed to direct its brightest minds and deepest investments. The path toward cyborg synthesis was not one Musk took alone. He was only one of a hypervisible indicator species for the consequences of fusing the economy ever more tightly with the digital world—a process that also destroyed the foundations of legacy political parties and fueled the emergence of disruptive new social movements on both the left and the right.

Chapter Five explores Musk's growing immersion in social media and his manipulation of its recursive loops for financial gain. This would be a world of memes and phantasms designed to create cycles of engagement and attachment. Chapter Six discusses his early experiments in AI and brain-computer interfaces, as well as the consequences of the pandemic. Chapter Seven narrates Musk's growing obsession with what he calls the "woke mind virus," alongside his efforts to dispel this infection: first by acquiring Twitter, then by founding his own AI company. Finally, Chapter Eight follows Musk's crusade against wokeness into the White House, in particular through the Department of Government Efficiency (DOGE) initiative that he pursued during the first months of Trump's second term.

If Muskism promises sovereignty through technology, the second part of our book illustrates the evolution of that promise. The next phase would be secured not only through Gigafactories and Starbases but through controlling the flow of data between minds and machines to prevent ideological contagion. As the world grew more digitized, more of our shared reality would be mediated by software. As artificial intelligence became more capable, human decision-making would evanesce. Life's messy contingencies would give way to the clean predictabilities of code.

5

Attention Alchemy

In 2004, the tech guru Tim O'Reilly started hosting a new conference in downtown San Francisco. Investors and entrepreneurs packed into a hotel for three days of talks featuring luminaries like Marc Andreessen and Jeff Bezos. The theme was "Web 2.0," a term that O'Reilly had recently begun using.[1] If Web 1.0 referred to the dot-com era that had come to a close with the bursting of the bubble, Web 2.0 described the internet economy that arose in the aftermath. It could be captured in a single word: platform.

For O'Reilly, a platform was an "architecture of participation." Instead of passively consuming a website, users would be enlisted as its co-creators. Their content, activity, and interactions would power the next era of the web.[2] As John Battelle, a co-founding editor of *Wired*, explained in the conference's opening presentation, Web 2.0 was about "building your business by letting your customers build your business."[3]

These concepts weren't completely novel. The notion that the internet was a uniquely interactive and participatory medium had been a truism of dot-com thinking back in the 1990s. When, in 2005, O'Reilly wrote about using "the power of the web to harness collective intelligence," he could have been writing ten years earlier.[4] What had changed was that Silicon Valley had finally begun to figure out a business model that converted these qualities of the internet into a revenue stream.

The formula, first perfected by Google, was what the scholar Shoshana Zuboff later termed "surveillance capitalism."[5] Maximize user engagement in order to generate as much data as possible about your users. Then monetize the data, primarily by selling targeted ads. This strategy enabled Silicon Valley to overcome the challenges of commercialization that had bedeviled the dot-commers and start making real money from the internet. Social media was just getting off the ground when O'Reilly held his first Web 2.0 conference: Facebook was founded in 2004, YouTube in 2005, Twitter in 2006. This new generation of "social" startups would define Web 2.0 more than any other.

Web 2.0 wasn't just built on clever technology—it was a product of monetary policy. After the 2007–8 financial crisis, the Federal Reserve slashed interest rates below 1 percent and kept them there for nearly a decade. It also printed money to buy government bonds and other securities. Investors looking for higher returns gravitated to Silicon Valley. As the scholar Nick Srnicek points out, this was the political economy that enabled the rise of the platforms.[6] If the government had produced the conditions for Web 1.0 by funding the creation of the internet, it did the same for Web 2.0 by engineering a long period of ultra-cheap money.

In 1998, Elon Musk had predicted that the internet would become "the superset of all media." The arrival of the platform age made good on this promise: U.S. newspaper advertising revenue fell by nearly 63 percent from 2006 to 2016, while online advertising revenue grew by more than 300 percent over the same period, most of which was captured by Google and Facebook.[7] Google cracked the profitability problem early—its revenue rose an astonishing 3,590 percent less than four years after figuring out how to monetize user data.[8] The same newspapers that Musk had hoped to disrupt with Zip2 were now being swallowed up by the internet superset. Music, movies, and all other media would follow.

Eventually, it would swallow him up as well. In particular, he would become obsessed with social media. He would become extremely online, an incurable poster. From the start, social media platforms were designed to be addictive. Most founders were wise enough to keep their distance. Musk, by contrast, threw himself in. He did so not only for the psychic rewards but for the financial ones. Social media, he discovered, could be a powerful engine of capital accumulation.

Musk had long deployed the method of financial fabulism to raise money for his companies. Drawing on a lesson he had first learned in Silicon Valley during the dot-com boom, he spun persuasive science fictions to win the confidence of his investors. More broadly, his career had always involved the judicious cultivation of social capital. In public he projected the persona of a playboy nerd, perhaps best captured by his famous cameo in the 2010 superhero movie *Iron Man 2*. The film's depiction of Tony Stark, the billionaire protagonist, was modeled partly on Musk.

With social media, Musk found a new tool for cultivating social capital. But its dynamics were distinct. The internet superset, as he had noted in 1998, absorbed older media while adding an overlay of interactivity. Social media was a space of conversation and contestation, love fests and screaming matches. Users didn't just interact; they competed for one another's attention. Viral marketing was a term from the 1990s, but social media gave it concreteness: "going viral" came to describe a specific kind of online popularity, precisely measurable through the number of reactions and shares. Facebook introduced the thumbs-up "Like" button in 2009. Other platforms followed. In 2016, Facebook expanded the Like button to include five additional emotions: "love," "haha," "wow," "sad" and "angry." What the sociologist William Davies calls "the reaction economy" became central to the real economy.⁹

On Twitter, Musk would become adept at going viral. More importantly, he developed a way to leverage virality into value by honing a new technique: attention alchemy. Attention alchemy involved taking a cue from the platform business model: the goal was to maximize and monetize engagement. For Musk, monetization didn't come from selling ads but from inflating the prices of financial assets. He drove engagement with hype and humor and then turned engagement into wealth, especially through the meme and its monetized offspring: the meme coin and the meme stock.

In 1994, William Gibson had worried that the internet would become an "infomall." The public network would be replaced with a privatized marketplace. By the 2010s, the internet had been remade into a marketplace where attention itself had become the primary currency. If Web 2.0 was about exploiting the "collective intelligence" of the crowd in a time of low interest rates, attention alchemy was a method ideally suited to it. Loose monetary policy converged with the deepening fusion of tech and finance to produce a historical moment where one man could move markets with memes.

Attention alchemy would be more than just another module of Muskism. It would also help bring about a new stage in Muskism's evolution. By embracing Twitter, and finding the financial advantage in its feedback function, Musk grew increasingly persuaded that the boundary between human beings and computers, between the organic and the digital, was dissolving, and that success required embracing this dissolution rather than fighting it. The idea recalled the image of the mech from the animated shows of Musk's youth, where human pilots harmonized with intelligent hardware through a neural uplink. They did so in order to defend their homelands from apocalyptic dangers. Muskism would come to be ruled by a comparable conviction: to be victorious, you must merge with the machine and never, ever log off.

Wokes and trolls

One way to think about the difference between Web 2.0 and Web 1.0 is that Web 2.0 fulfilled the promises that Web 1.0 made but couldn't keep. These concerned not only the internet's commercial potential—loudly proclaimed by the dot-coms but never quite consummated—but its political potential as well. The techno-utopians of the 1990s saw the growth of computer networks as a democratizing force. "If we learn anything from the collapse of the Berlin Wall and the fall of the governments in Eastern Europe," announced President Bill Clinton in 1993, "[it's that] even a totally controlled society cannot resist the winds of change that economics and technology and information flow have imposed in this world of ours."[10] Technology would advance the cause of political freedom by making information flow freely.

By the late 2000s, this judgment seemed to be coming true. The "architecture of participation" celebrated by O'Reilly as a pillar of the new platform era was becoming a powerful tool for activism all around the world. Pundits began talking about Twitter Revolutions and Facebook Revolutions.[11] In 2011, uprisings from Cairo to Madrid to Athens used social media to coordinate and broadcast their struggles to a watching world. In

July of that year, *Adbusters*—the Canadian magazine that popularized the idea of "culture jamming" in the 1990s—launched Occupy Wall Street with a striking image: a ballerina poised atop the Charging Bull statue in Manhattan's financial district. "What is our one demand?" the poster asked, alongside the instruction: "Bring tent."[12] Many people did. The encampment in Zuccotti Park began in September and lasted for months. Copycat movements spread across the country and the world. In December, *TIME* named "The Protester" its Person of the Year. Among the portraits included in the issue was one of the Tunisian blogger Lina Ben Mhenni.[13] She looked steadily into the camera, an Apple laptop covered in stickers tucked under her arm. "Dictatorship can't last," the text read. Social media looked like a tool that was shifting the balance of power: weakening governments while strengthening popular self-determination.

The political hashtag was a prime artifact of the moment. Every movement needed one. If there wasn't one, foreign journalists would supply them—as with the #JasmineRevolution that led to the ousting of Tunisia's longtime dictator in 2011.[14] The hash symbol turned phrases into search terms and rallying cries: internet functionality became political expression. #LoveWins spiked after the Supreme Court legalized same-sex marriage nationwide, #JeSuisCharlie after the attack on the magazine's offices in Paris. #BlackLivesMatter, first used in 2013 after the acquittal of George Zimmerman for the killing of Trayvon Martin, surged again after the police killing of Michael Brown in Ferguson, Missouri in 2014.

By 2016, these political expressions had migrated from the margins to center stage. At the Code Conference in San Francisco that summer, Twitter CEO Jack Dorsey appeared beside Black Lives Matter activist DeRay Mckesson wearing a T-shirt that read "#StayWoke." (The shirts had been made by Black employees at the company after the police killing of Michael Brown in 2014). Everyone in the crowd received their own.[15] "Woke" was a word with deep roots in Black culture: as early as 1938, the folk singer Lead Belly had urged his listeners to "stay woke" in a song about the Scottsboro Boys. Staying woke meant being alert to injustice, and to the likelihood of police and white vigilante violence.[16] Dorsey coopted the term for his own platform. Twitter offered a way to revise the narratives of mainstream journalists and deliver a more authentic perspective on current events.[17] For him, scrolling and posting were the way to remain woke in the contemporary world.

In the coming years, the term would become a catch-all for progressive politics, as amplified through the megaphone of social media. Yet the woke social network of #HandsUpDontShoot and #MeToo was always shadowed by its obscene doppelgänger: the internet of trolls. In 2003, a year before Facebook was founded, another tech-savvy teenager launched a website called 4chan. It was a simple message board modeled on a Japanese predecessor named 2chan, and its founder was a fifteen-year-old student from New York City named Christopher "moot" Poole. 4chan was a networked public of pubescent humor and smut, shot through with homophobic, racist, sexist, and ableist slurs. It also incubated many mainstays of internet culture, such as the meme.

The term "meme" was first coined by the evolutionary biologist Richard Dawkins in the final chapter of his 1976 book *The Selfish Gene*, where he described it as a "unit of cultural transmission," such as an idea, song, or style of clothing.[18] By the 2000s, the term had evolved far from its original context to mean a combination of text and image shared online. As the media scholar Limor Shifman explained in an influential book, memes were defined by two aspects: mimicry and remix.[19] Early examples distributed on 4chan included "LOLcats" featuring images of cats overlaid by text written in intentionally fractured language called "LOLspeak" such as the iconic "I Can Has Cheezburger?" The visual grammar was straightforward: a photo, bold white Impact font, a setup line on top, a punchline below.

4chan also popularized the provocative online style known as trolling.[20] The trolls took it as their mandate to playfully upend the efforts at collective consensus creation embodied by hashtag activism. As the scholar Whitney Phillips has shown, the etymology of the term "troll" is disputed.[21] Some claim that when the term was first used on online discussion boards in the 1990s, it referred to the hirsute creature of folklore. More plausible, however, is that it relates to the fishing term—trolling as trawling. In practice, it has come to mean both: the person who incites a reaction is rewarded with the currency of today's internet—namely, attention—and in doing so, becomes something grotesque and troublesome. If "woke" users posted for justice, trolls posted for laughs—or "lulz," described by journalist Mattathias Schwartz as "the joy of disrupting another's emotional equilibrium."[22] The troll was the fly in the ointment of the aspirational public sphere.

4chan first became known to a wider audience in 2016, the same year that Dorsey took the stage in San Francisco in a #StayWoke T-shirt. That spring and summer, as Hillary Clinton and Donald Trump crisscrossed the country for their presidential campaigns, scandalized Americans learned that online trolls were referring to Trump as "God Emperor" and had appropriated a web comic character called Pepe to spread anti-Semitic messages and target journalists with surnames assumed to be Jewish.[23] On her campaign website, Clinton published a somber warning that the "cartoon frog is more sinister than you might realize."[24] Delighted by the traction, the trolls doubled down.

They had already scored a major victory when Trump retweeted an image of himself as Pepe at the podium.[25]

That fall, #MAGA eclipsed #BlackLivesMatter as the most popular political hashtag on Twitter.[26] If this was a Twitter Revolution, it was not the kind that most journalists had wanted. With Trump's surprise election in November 2016, the tenor of public discussion about the internet changed. The new keywords were "misinformation/disinformation," "algorithmic manipulation," and "online harassment."[27] Concerns about the role of Russian influence operations on social media became a staple of mainstream discourse. Mark Zuckerberg was summoned to Capitol Hill to testify about whether Facebook had helped skew the election. In the UK, there were intimations that a company called Cambridge Analytica had used dark digital arts to help tilt the Brexit vote toward Leave.

"Sharing" had been the rallying cry of Web 2.0. The term's usage grew from the mid-2000s to the mid-2010s but, according to the scholar Nicholas A. John, it fell off sharply after that point.[28] A reasonable conclusion is that people were no longer fooled. They may have been "sharing" their lives, but the companies were selling them. In a 2018 congressional hearing, an elderly senator asked Zuckerberg how Facebook makes money. Zuckerberg smirked and replied: "Senator, we run ads."[29]

By the end of the 2010s, the reputation of social media had entered a twilight stage—what one academic paper called social media's "midlife crisis."[30] More people than ever used the platforms—especially as smartphones reached near-total saturation among adults—but public opinion had soured. Sixty-four percent of Americans believed that social media was making things worse rather than better.[31] Rather than social media activism, the talk was about social media addiction. Enthusiasm had curdled into anxiety.

In 2019, Shoshana Zuboff's bestselling book *The Age of Surveillance Capitalism* codified the new common sense. The data-driven advertising model of Web 2.0—once hailed as a brilliant business innovation that had turned the internet into a powerhouse of global capitalism—was now seen as a monster that devoured our privacy and shredded our democratic institutions.[32] "What is at stake," she wrote in the alarmist register that would become de rigueur after 2020, "is the human expectation of sovereignty over one's own life and authorship of one's own experience."[33]

Troll capitalism

It was a strange time to decide to become extremely online. Yet that is exactly what Musk did. He first tweeted in 2010 but remained, at most, a casual Twitter user. Starting in 2015, this began to change. That year, he tweeted 617 times—an increase of 226 percent from the previous year. It marked the onset of Musk's tumbling into the social media fray. In 2017, he tweeted 1,162 times—an average of three posts per day. By 2018, his annual tweet count had hit 2,288. By 2024, he was posting an average of sixty times a day and sometimes as often as forty times an hour.[34]

For someone so deeply involved in the heavy materialities of Tesla and SpaceX—the extraction of minerals, the calibration of machines, the reshaping of human labor on the factory floor—Twitter held the allure of immediacy. Here, there was no lithium to worry about. No need to strategize around the Chinese Communist Party or fend off competitors. Just *lols* with friends, with tens of millions of people watching. The casino logic of social media suited Musk. At any hour, he could drop another coin in the slot machine and hope to go viral.

Immersion in social media was not just welcome stress relief, however. It also became an integral part of Musk's business strategy. He had long taken negative coverage personally—he once compared it to being "pistol-whipped"—and insisted that journalists who wrote critically about him or his companies had ulterior motives for doing so.[35] On social media, he could hit back at the headlines and take control of the narrative. "With Twitter, we can talk directly to the people," he told a SpaceX executive in 2016.[36] By disintermediating the media's traditional gatekeepers, he could draw investors deeper into his reality and stoke the hype around new products and features.

His communication style had always been proleptic. The logic of financial fabulism treated imagined futures as already underway, allowing speculative claims to generate market effects before the underlying technology had matured. "Like the Soviet state dangling the promise of a radiant future in front of its tired citizens," the critic Phil Jones observed, "Musk's success is sustained by predictions of a technological sublime that's only ever another decade away."[37]

On Twitter, however, such predictions could produce financial effects instantaneously. In 2018, Musk tweeted, "Am considering taking Tesla private at $420. Funding secured."[38] The number was a weed joke, but investors took him seriously: Tesla stock jumped 11 percent. A civil suit from the SEC followed. Both he and Tesla agreed to pay $20 million in fines.[39] Still, the incident offered early proof of how quickly his tweets could move markets. In May 2020, he posted "stock price too high imo" and Tesla dropped as much as 12 percent.[40] In January 2021, he added "#bitcoin" to his Twitter bio and the cryptocurrency jumped 20 percent within an hour.[41] This was attention alchemy at work. As the journalist Marco D'Eramo observed, Musk's followers were his "real capital."[42]

What even the sharpest observers tended to miss, however, was the centrality of humor—even bad humor—to the model. Musk had traditionally been the entrepreneur who *made stuff*. But from the mid-2010s onward, his success would also be driven by something sillier and more ephemeral: online jokes. Tonally, these jokes drew from the troll hothouses of 4chan, Reddit, and gaming circles. At times, they edged into toxic territory. In 2018, after British cave explorer Vernon Unsworth criticized Musk's unhelpful efforts at rescuing twelve kids stuck in a cave in Thailand, Musk jokingly called him a "pedo guy" on Twitter.[43] When Musk was taken to trial, he lashed out at the press. After winning the case, he fired his public relations firm.[44] He was already breaking up with the legacy media and edging closer to the internet of trolls.[45] Social networks were his message machines and, as he had shown, could be money machines too.

Meme city

In the 2010s, what Nick Srnicek calls the "new conglomerates" were emerging. Facebook bought Instagram (2012) and WhatsApp (2014), Google became Alphabet (2015), and Microsoft bought LinkedIn (2016).[46] For decades, firms were getting smaller; now they were getting bigger.[47] As before, Muskism coalesced with these new elements of political economy as they came online. These had included the expansion (and privatization) of the national security state, the surge of green capitalism, the growing dominance of Silicon Valley, and the rise of China. By the late 2010s, Muskism was also manifesting in the meme economy of troll capitalism.

In February 2019, Musk appeared on the stream of the gaming influencer PewDiePie with Justin Roiland, co-creator of the animated sitcom *Rick and Morty*—one of Musk's favorites.[48] Roiland asked Musk to review and rate some memes. At one point, Roiland showed Musk a meme made from a photograph of his face. "Is that trippy?" Roiland asked. "Or have you gotten over that?" Musk shrugged. By that point, it felt natural. He had become the hinge between the chaotic churn of internet subcultures and the buttoned-down world of capital:

> MUSK: The past couple years have been—particularly last year—it was meme city.
> ROILAND: Meme city.
> MUSK: Yeah. Meme city.

Or, as he put it five months later on Twitter, "I am become meme."[49] It would be a running joke for Musk, riffing on J. Robert Oppenheimer's famous invocation of the Bhagavad Gita after witnessing the first detonation of a nuclear weapon in 1945: "Now I am become Death, the destroyer of worlds." If the goal of coders in 1990s Silicon Valley, where Musk cut his teeth, was to enter a flow state, a kind of symbiosis with the computer, Musk's repeated statements that he had "become meme" suggested something similar. A surrender to the timeline. A willingness to let the medium dictate the message.

Becoming meme was a subjectivity that invited participation. Musk spends most of his time not broadcasting to his audience but reacting to it. Roughly three quarters of his posts are replies—not declarations but engagements.[50] As the journalist Nick Bilton has observed, it's not common behavior for a Fortune 500 CEO to block journalists he dislikes.[51] But it's even more rare for them to stay up all night responding to random accounts online. Unlike traditional corporate titans, Musk does not simply speak *at* the internet; he speaks *with* it.

In other words, although it is possible to see Musk as an egomaniac or someone afflicted with Main Character Syndrome, it is also intriguing to frame him a bit differently: as someone whose willingness to enter into the medium as a peer, to play the game alongside others, is part of his power.

To be a meme is to be modular, reusable—part of a circular economy that is ever-growing. This too is part of Musk's playbook. He finds certain things "concerning," lets others connect the dots, and blesses the crowd's narrative. In a media system governed by virality, this is an optimized form of influence. The more absurd the content, the more persuasive the effect. If people would buy this, what wouldn't they buy?

Musk was attuned to the prevailing sense of disenchantment with the internet, one that no longer believed in digital spaces as revived public spheres or engines of social justice, but as sites of doomscrolling, sadism, and schadenfreude. Twitter super-users referred to the platform as "the hellsite." At its most productive, it was something simpler, rawer, and more seductive: a place to make money. Here comes everybody—to make the line go up.

Doge tales

By the 2010s, as memes outgrew their message board origins to become a mass phenomenon on social media, their aesthetic evolved. Comic Sans in rainbow colors replaced Impact in white. Text fragments of different sizes were scattered across the image. The breakout star of the new style was a wide-eyed Shiba Inu dog from Japan, soon to be known as Doge.

The history of Doge reveals how obscure online artifacts migrate into the mainstream. The term "doge" first surfaced in a 2005 episode of a web series featuring two sock puppets in an office setting. One puppet keeps repeating that he is the other's "d-o-g-e," a mispronunciation played for laughs. Five years later, a Japanese kindergarten teacher posted photos of her photogenic Shiba Inu. By 2013, Reddit and Tumblr users had combined one of those images with the misspelled "doge"—and the meme was born. It soon acquired captions in broken English: "so scare," "much wow," "very art."[52]

In principle, part of the charm of memes was their utter worthlessness. You paid only in the laugh—or in the brief flicker of recognition from another user. In 2010, 4chan's founder appeared onstage at the annual TED conference to explain that the commercial model of 4chan was that "there really isn't one."[53] In 2013, the tension between carnivalesque play and incipient commodification was put on display when two engineers launched Dogecoin, calling it a "peer-to-peer parody cryptocurrency" based on the Shiba Inu meme. Its website was plastered with floating phrases like "such currency," "how money," and "v rich."[54]

The punchline was the disjuncture between a goofy mascot and the speculative logic of digital wealth. Dogecoin was born as a joke, but the joke turned out to be functional. It rode the wave of a still-niche interest in cryptocurrency, a system enabled by blockchain technology. Bitcoin, the original blockchain-based currency, had launched in 2009 and was worth about $830 per coin when Dogecoin emerged. (By 2025, Bitcoin would trade for well over $100,000.)

The creators of Dogecoin wanted to highlight the Dada-like uselessness of the meme as money. A comic that circulated on a Tumblr page called "Fuck Yeah Dogecoin" captured the absurdity. A stick figure watches a breaking news alert: "With the collapse of the dollar, the government has endorsed an alternate currency. Your monetary worth is now determined by the number of funny pictures saved to your hard drive." The caption read: "I have been preparing for this moment my whole life."[55] What could be more ridiculous than taking this kind of joke seriously?

Then, just a few years later, something strange happened. Dogecoin started to be worth something. One reason was because Musk started to post about it. "Dogecoin might be my fav cryptocurrency," he tweeted in April 2019, "It's pretty cool."[56] "Dogecoin is the people's crypto," he posted in February 2021.[57] In January and February of that year, Musk unleashed a string of Doge-themed posts.[58] A *Vogue* cover doctored to read "Dogue." A rocket blasting into space, labeled simply "Doge." A Shiba Inu in a spacesuit, posing triumphantly on the moon. A Photoshopped still from *The Lion King*, with Musk in the role of Rafiki, holding up the Shiba Inu as if it were Simba. Each post added fuel to the fire. Each meme, a microdose of market motion.

Musk said that what he liked about Dogecoin was its "humor & irreverence."[59] But the coin also disclosed a deeper feature of Muskism. The cryptocurrency, as Bloomberg columnist Matt Levine put it, could be seen as a "tradable electronic token representing the value of Elon Musk's attention."[60] The logic was straightforward: if Musk tweeted about Dogecoin, the price jumped. The possibility of pumping that value—and then dumping it through strategic sales—was enormous. But was Musk doing that? A legal case attempting to prove that he was manipulating the market failed.[61] Meme coins were in a gray zone between a gag and a business, between a line of code and a unit of value, between a provocation and a proposition. A zone in which Muskism thrives.

Stonks!

In March 2019, a single Dogecoin was worth around a quarter of a cent. Two years later, it was worth about five cents—an increase of more than 2,000 percent. If Musk was one reason for this development, the other was Covid-19. The global pandemic that began in March 2020 would kill more than a million Americans. It would also supercharge the political economy of Web 2.0 and make the material conditions for attention alchemy even more propitious.

The initial fallout of the pandemic was severe: the U.S. economy contracted sharply, the stock market crashed, and unemployment spiked. The Federal Reserve responded by resurrecting its firefighting toolkit from the 2007–8 financial crisis. It cut interest rates and kicked off a massive round of quantitative easing by purchasing hundreds of billions of dollars' worth of financial assets, among other measures.[62] Congress also stepped in to provide relief through a series of bills that pumped trillions into the economy, including by expanding unemployment benefits and sending people stimulus checks.

At the same time, Americans began spending more hours online. Kept indoors by lockdown measures, they worked from home, attended virtual classes, shopped, socialized, and wasted time. Together, these two factors laid the foundation for a new wave of tech–finance fusion. The government's monetary and fiscal interventions made capital even more abundant, encouraging investors to place riskier bets. Silicon Valley benefited from this macroeconomic environment, as well as from the increased digitization of daily life. Big tech stocks boomed while work-from-home darlings like Zoom, Peloton, and DocuSign saw their valuations soar. Retail investing also exploded, as people stuck at home chased instant riches on brokerage platforms like Robinhood, where trading surged 139 percent in the second quarter of 2020.[63] Among the assets they plowed their money into was Dogecoin.

By this point, Musk had spent years splicing himself into Twitter's memetic circuits. The pandemic increased the supply of attention to monetize and the liquidity with which to do so. He responded not only by pumping Dogecoin but also by joining the "meme stock" craze that took off in early 2021. The canonical example was GameStop, the strip mall video game chain whose stock was rescued from oblivion by retail investors coordinating online. "Gamestonk!!" Musk tweeted in January 2021, linking to WallStreetBets, the Reddit forum where the action was unfolding.[64] His misspelling was itself an inside joke, a mutated descendant of LOLspeak. "I am become meme, Destroyer of shorts," he tweeted the following month, a reference to the efforts of the WallStreetBets community to inflict large losses on hedge funds that were shorting GameStop.[65]

WallStreetBets was the epitome of troll capitalism. "Like 4chan found a Bloomberg terminal," its tagline read. If trolling was a way to harvest attention, then meme stocks proved that trolling could become big business. One of the best diagnoses of the shift came from an unlikely source: Jackson Palmer, an inventor of Dogecoin. "In this system of griftonomics, hypercapitalism, rentier capitalism," he told an interviewer in May 2022, "increasingly people are doing nothing but making money off doing nothing." As for Musk, Palmer added, he "was and always will be a grifter but the world loves grifters. They love the idea that they may also be a billionaire one day, and that's the dream he's selling."[66]

There are few things more American than the fantasy of instant riches. What Musk offered was the possibility of fulfilling this fantasy through the power of memes. "Who controls the memes, controls the Universe," he posted in June 2020, an ironic reformulation of a well-known line from the science-fiction epic *Dune*.[67] It seemed like a fully dematerialized view of political economy, far from the world of physics that he had always claimed fealty to as an engineer and which had earned him grudging admiration even from people who found him personally unpleasant. It suggested that, for Musk, the internet was becoming the primary layer of reality—that, contrary to the adage that "Twitter isn't real life," Twitter was becoming realer, and more consequential, than real life.

What made such a belief plausible was the fact that meme coins and meme stocks had the capacity to enrich Musk beyond his wildest dreams. In the words of Matt Levine, Musk had a magic lamp in his hands. "You can just whisper 'price go up' or 'price go down' into the lamp, and it will happen instantly," Levine noted. "You are the only person who can do this, and you can do it as often as you want." It's "as close to a free-money perpetual motion machine as you'll ever see in finance."[68]

This development wasn't entirely without precedent. Musk had "invented and perfected meme stocks long before anyone ever thought about GameStop," observed Levine.[69] The first meme stock wasn't GameStop but Tesla. Musk had long been inflating the valuation of the car maker through posting. The arrival of the pandemic pushed Tesla to new heights: its share price rose by more than 740 percent in 2020, becoming the world's most valuable car company. By 2021, Musk's Tesla holdings had made him the richest man in the world. His process of "becoming meme" had culminated with his transformation into what the journalist Charlie Warzel called a "human meme stock."[70] The consequences for Muskism would be profound.

6

Cybernetic Collectives

Artificial intelligence has been an aspiration since the 1950s. At the height of the Cold War, a group of computer scientists persuaded the federal government to fund efforts at universities across the country to build intelligent machines. The results were underwhelming. Then, in the early 2010s, a new generation of AI systems powered by a data-processing architecture called a neural network began to make rapid progress on stubborn problems, like getting a computer to recognize objects in images.[1] Advances in AI helped speed the rise of the platforms, as Google, Facebook, and others adopted increasingly sophisticated models to optimize their operations. Tesla, too, joined the fray: in 2016, the company started introducing neural networks into its cars' self-driving software.[2]

As AI became more powerful in the 2010s, however, it rekindled age-old fears about evil robots. Elon Musk shared these fears. In fact, they were his motivation for co-founding OpenAI in 2015 with Sam Altman and several others—the same year that he ramped up his Twitter presence. He would remain co-chair of OpenAI's board until his departure in 2018. His objective, he later explained, "was to increase the probability that AI would develop in a safe way that would be beneficial to humanity."[3] In particular, he worried that a "superintelligent" AI could enslave or exterminate the human race in the near future, a scenario informed by his reading of the philosopher Nick Bostrom, whose 2014 book *Superintelligence* circulated widely in the AI world. Musk praised the book and was thanked in its acknowledgements.[4] The same year, he told a symposium at MIT that AI was "our biggest existential threat."[5] Later, when the AI researcher Timnit Gebru approached Musk at a Stanford event and asked why he considered AI more dangerous than climate change, he clarified his reasoning. "Climate change is bad, but it's not going to kill everyone," he said. "AI could render humanity extinct."[6]

Musk's view of AI is often misunderstood. His frequent invocations of *Terminator*-style scenarios have led him to be labeled an AI pessimist. In 2015, he was even nominated (absurdly) for Luddite of the Year.[7] This got things backwards. Musk's solution to the risks of AI was not less technology, but more. The threat would be overcome not by resisting integration, but by accelerating it. As he put it in a 2016 conversation with Altman, the answer was to "merge with AI." If "you become an AI–human symbiote," he said, "we don't have to worry about some sort of evil dictator AI because we are the AI collectively."[8] What Musk called the "democratization of AI power" meant the scaling of this synthesis as a safeguard against artificial tyranny.[9]

The first step was to recognize how entangled we already were with our machines. The human mind, as he saw it, is composed of three layers: the primary "animal brain," or limbic system, which governs emotion and instinct; the cortical layer, seat of reason and deliberation; and a tertiary layer—an expanding array of "digital tools" that extends our capacities.[10] "We are effectively already a human–machine collective symbiote," he told Altman at a *Vanity Fair* event in 2015, "We're like a giant cyborg. That's actually what society is today."[11]

The most important site of cyborg symbiosis was social media. "Facebook and Twitter and Instagram and all these social networks—they're giant cybernetic collectives," Musk told the podcaster Joe Rogan in 2018. They didn't just let people collectivize their thoughts but, more importantly, their feelings. The "success of these online systems," Musk argued, is "a function of how much limbic resonance they're able to achieve with people." Virality was driven by emotion. "The more limbic resonance, the more engagement."[12]

What made these collectives *cybernetic* was the fact they included computers as well as humans. And these computers were, in fact, learning from the humans. AI systems based on neural networks are trained to perform a particular task by finding patterns in large quantities of data. On the platforms, users supplied this data through their activity. "We're all collectively programming the AI," Musk explained.[13]

Gradually, this process would result in smarter and smarter AI. "The percentage of intelligence that is not human is increasing, and eventually we will represent a very small percentage of intelligence," said Musk. The ultimate legacy of the cybernetic collectives of social media would be humanity training its replacement. In a computer, a bootloader is a special program that helps initialize the system. Humanity, Musk told Rogan, was becoming "the biological bootloader for AI."[14]

But there was an interesting wrinkle to this theory. If our online interactions were fueled more by emotion than reason, then the AI systems that we were collectively programming would reflect that. The AI that learned from observing our behavior in the cybernetic collective would become "our id writ large," Musk said.[15] This was a view of advanced AI not merely as "superintelligence" but as an algorithmic embodiment of our combined impulses and instincts.

These quotes offer a revealing glimpse into Musk's shifting outlook in the mid to late 2010s. His integration into Twitter's feedback loops had given him not just a new way to get rich by alchemizing attention into value but an all-encompassing vision of a cyborg future. In the course of "becoming meme" he was coming to understand social media as a cybernetic collective that was helping humanity evolve into something posthuman.

With Web 2.0, Silicon Valley had discovered the value of participation. By building platforms that encouraged users to interact, companies made a fortune by monetizing the data that resulted. The next era of the internet, Musk suggested, would involve using that data to train AI models. We would become AI's organic symbiote and "biological bootloader." The wisdom of the crowd would be enlisted to tutor a new species of smart machines.

One implication was that social media had immense importance for the future of the human race. If social media were the primary site of cyborg symbiosis, then a platform like Twitter was more than a place to crack jokes, troll rivals, or pump crypto and stocks. It was a place where the perils of superintelligence could be neutralized by dissolving ourselves into data. If we didn't become AI, AI would eliminate us.

This mandate would define the new stage of Muskism that was becoming visible by the late 2010s. Karl Marx had once described the imperative to accumulate capital as the core commitment of a capitalist system. "Accumulate, accumulate! That is Moses and the prophets!" he wrote.[16] The Muskist imperative would be to accelerate the fusion of flesh and code, of meatspace and memesphere, in order to build a better cybernetic collective.

It wasn't enough to simply speed up the process, however. The cybernetic collective was vulnerable to infection. It could be compromised, just like any biological or digital system. This possibility, and the terror it provoked, would guide Musk's major undertakings of subsequent years, from his acquisition of Twitter to his foray into the second Trump administration. Muskism would be entranced by the dream of the cyborg and haunted by the nightmare of its potential contamination.

Becoming bot

In 2016, the year after co-founding OpenAI, Musk started Neuralink. The company planned to create a brain–computer interface that let people control digital devices with their minds. Neuralink shared office space with OpenAI in San Francisco.[17] The two might seem like dissimilar ventures, but they had a common mission: to prevent the AI apocalypse.

"This is going to sound pretty weird," Musk announced in Neuralink's first public presentation in 2019, but the company aimed to "achieve a sort of symbiosis with artificial intelligence."[18] The brain implant, which he claimed would eventually be inserted into "hundreds of millions" of people, would let users connect their minds directly to the internet, accelerating the man–machine merger that was already underway in the cybernetic collectives of social media.[19]

But social media had a limitation. If Twitter made us into cyborgs, it did so in a primitive way. The bottleneck, Musk believed, was the interface. Bandwidth was throttled by ten clumsy fingers, tapping on a keyboard. "The communication rate between you and the cybernetic extension of yourself is slow," Musk warned. It's "like a tiny straw . . . we need to make that tiny straw like a giant river."[20] Neuralink promised to vastly increase the bandwidth between digital and biological systems. One of Musk's favorite films is *Ghost in the Shell*, an anime from 1995 set in 2029. It features characters with cyborg hands that can type not only individual letters but *chords*—thus massively increasing the speed of data entry. This was something like how Musk hoped Neuralink would work. And such a breakthrough would, in turn, reduce the risk of a superintelligent AI killing us all by facilitating our unification with the machine. If one goal of Neuralink was "solving a bunch of brain-related diseases," Musk said, the other was "the mitigation of the existential threat of AI."[21]

Neuralink bridged the two worlds described in the first half of this book: the software of Silicon Valley and the hardware of rockets and cars. It also drew on a familiar theme of Muskism: state symbiosis. In 1973, the computer scientist Jacques Vidal coined the term "brain–computer interface" (BCI). The following year, DARPA—the same agency that financed the creation of the internet and provided SpaceX with one of its first federal contracts—launched a Cybernetics Technology Division.[22] The concept of "network-centric warfare" that had been so important to SpaceX's early success extended into fantasies of neurally enhanced "supersoldiers" operating without sleep and wired for quicker reaction times.[23]

Another impetus behind Pentagon investment in brain–computer interfaces was the desire to improve prosthetics for the soldiers returning from Iraq and Afghanistan with missing limbs.[24] In 2006, DARPA announced a four-year plan to build a robotic arm "directly controlled by neural signals."[25] In 2014, it created a Biological Technologies Office, which invested more than $200 million in neurotechnology over two years.[26]

Musk's cyborg talk was not the esoteric speculation of a lone visionary, then, but the rhetoric of an entrepreneur entering a crowded field. While Neuralink did not receive direct federal funding, it benefited from decades of publicly funded research. Admittedly, Musk's motivation was less typical: he hoped to use Neuralink not just to help people with disabilities but to catalyze the cybernetic synergy required to forestall the threat of human extinction at the hands of AI. Here, too, however, he was far from alone. His framing drew on author-inventors like Kevin Warwick and Ray Kurzweil who promoted "transhumanist" hybridization, especially in the latter's influential book *The Singularity Is Near* (2005).[27] In 2010, Kurzweil proposed a public competition to build brain–computer interfaces that would put artificial intelligence "inside our bodies and brains."[28] And in 2001, no less a luminary than Stephen Hawking had urged the development of "technologies that make possible a direct connection between brain and computer, so that artificial brains contribute to human intelligence rather than opposing it."[29]

This was exactly what Neuralink hoped to do. It was quite literally an attempt to create a cyborg. In this new stage of Muskism, the mech would be more than a metaphor. The ambition was not simply to build rockets and cars but to rebuild humanity itself. The logical conclusion of "never log off" was to hardwire the connection permanently into the human brain. Muskism hoped to create a species capable of surviving its own technological creations by becoming indistinguishable from them.

In January 2024, a quadriplegic named Noland Arbaugh became the first human recipient of a Neuralink chip. Two months later, he posted a tweet. "Twitter banned me because they thought I was a bot," it reads. "@X and @elonmusk reinstated me because I am."[30] By measuring the electrical signals in his brain and translating those signals into computer instructions, the Neuralink implant enabled Arbaugh to tweet just by thinking. The cybernetic collective had become that much more cybernetic.[31]

Mind control

The science behind Neuralink wasn't all that new. By the time Musk founded the company, progress had already been made on brain–computer interfaces that enabled people to manipulate a keyboard and a mouse.[32] But, as the journalist Jenny Kleeman pointed out, Musk was the first "entrepreneur whose explicit aim is to find a way to feed information *into* the brain, as well as receiving from it."[33] That is, Neuralink was distinguished by its desire to build an implant with a "write" function as well as a "read" function. In the cult 1999 movie *The Matrix*, the protagonist Neo learns kung fu in a matter of seconds by loading a program into his brain. Neuralink aimed to make this a reality.

Two-way information transfer was needed for the symbiosis with AI that Musk wanted to achieve. We couldn't merge with AI if the messages could only flow one way. Yet the prospect of a chip that could write data directly into the brain raised the specter of new risks. In 2019, Musk co-authored an article with the Neuralink team in the *Journal of Medical Internet Research*, laying out the company's "first steps toward a scalable high-bandwidth brain–machine interface system."[34] It was followed by a reply from three neuroscientists, who praised Neuralink's technical achievements but ended with a stark warning. "Among the undesired effects of brain–machine interfaces with electrodes implanted into the human brain," they wrote, "is the possibility that a government or a nongovernmental organization will control and manipulate the person's behavior not only through mass media, but also by directly sending commands to the brain."[35]

This was the idea of mind control. It had been a subject of phobic fascination for Americans since the Cold War, when movies like *The Manchurian Candidate* (1962) stirred public

CYBERNETIC COLLECTIVES

anxieties about Communist brainwashing. Brainwashing never actually worked: the CIA spent decades trying to figure it out with an illegal human experimentation program called MKUltra, without success.[36] But what if, Musk wondered, brain–computer interfaces like Neuralink would finally make it feasible? What if a mind could be programmed like a piece of software?

So far, Musk had conceived of the cybernetic collective as a solution to the existential threat of AI. With the arrival of the pandemic in 2020, however, he began to perceive a new set of dangers. By wiring everyone's brains together—as social media did metaphorically and as Musk believed Neuralink would do literally—the cybernetic collective might give bad actors a powerful set of tools with which to influence people's thoughts and feelings. In information security, every communication channel is also, potentially, an attack vector. If the pandemic benefited Musk in certain ways, it also provoked behavior that he had difficulty comprehending, behavior that ran counter to his interests. He came to interpret such behavior as the result of mind control, and as evidence that vigorous intervention was required to protect the psychic purity of the cybernetic collective.

On March 16, six Bay Area counties—including Alameda County, the location of Tesla's Fremont plant—issued a joint shelter-in-place order. Three days later, Governor Gavin Newsom extended the mandate to the rest of the state. While SpaceX's facility in Southern California was exempt on account of its "essential" status as a government contractor, the Tesla factory in Fremont had to suspend production.

What public health officials presented as collective responsibility appeared to Musk as coercion—an intrusion into the freedom of movement and production on which his empire depended. In his view, lockdowns compromised the autonomy of both individuals and enterprises. "FREE AMERICA NOW," he tweeted on April 29.[37] Later that day, he called the quarantine measures "fascist" on a Tesla earnings call.[38] President Trump sent a message of support over Twitter: "California should let Tesla & @elonmusk open the plant, NOW. It can be done Fast & Safely!" "Thank you!" Musk replied.[39] On May 11, he restarted production at the Fremont factory in defiance of the Alameda County order, daring local officials to arrest him.[40] The authorities caved, letting the factory reopen in exchange for promises to abide by minimal safety protocols like masking. The Tesla plant promptly became a hothouse of Covid-19 transmission, with around 450 cases recorded through the end of the year.[41]

Musk's anger was rooted in political economy. The company was uniquely vulnerable to a shutdown because its operations were relatively centralized. The value placed on vertical integration in Musk's industrial philosophy meant that people and processes were concentrated whenever possible. The Fremont factory still produced most of the world's Teslas. Closing it was painful, especially in the spring of 2020, just as the company had begun its first deliveries of its new SUV, which Musk hoped would help him broaden his customer base.

Yet the same qualities that made the shuttering of a single plant so existential would give the company a competitive edge when production resumed. Traditional automakers had been outsourcing for decades. They had grown reliant on increasingly baroque supply chains threaded across the world, which proved fragile in the face of pandemic-related disruptions. By contrast, Tesla's preference for vertical integration and shorter supply chains gave it greater resilience in a less stable world. Musk's decision to build a factory in Shanghai, which began producing vehicles in late 2019, also looked prescient. Building a manufacturing base inside China gave him access to the all-important Chinese market at a time when borders were hardening, not only on account of coronavirus control efforts but also due to the intensifying trade war with the United States.

These advantages paid dividends. In the second quarter of 2020, the first to register the full effect of the pandemic, Tesla reported only a 4.9 percent decline in deliveries from the same quarter in the previous year. Wall Street analysts expected a drop of 25 percent, in line with industry trends: Toyota, General Motors, and Ford all saw their sales fall more than 30 percent.[42] Investors rewarded Musk for outperforming his rivals. On July 1, 2020, Tesla became the most valuable car manufacturer in the world, with a market capitalization of $206 billion.[43]

Even so, Tesla was being rewarded far out of proportion to its actual revenue. Its price-to-earnings ratio, which measures the relationship between a company's stock price and its earnings per share, surpassed 1000 in December 2020. Toyota, which sold almost twenty times as many cars that year, enjoyed a price-to-earnings ratio of 13. This differential pointed to the importance of attention alchemy to Musk's fortunes. Tesla was, after all, a meme stock, one that benefited from the boom in online speculation triggered by the pandemic. If shelter-in-place orders endangered Musk's profit margins, another set of government policies—monetary and fiscal expansion—boosted them. The pandemic illuminated the inverted pyramid of Muskism: a narrow material base opening up into a vast virtual domain.

Superspreader events

Tesla's lockdown crisis was relatively brief. The Fremont factory shut down for only seven weeks. But for Musk, the experience highlighted a new source of risk that had to be reckoned with. What motivated the shelter-in-place orders, he believed, was panic. "The coronavirus panic is dumb," he tweeted in early March 2020, his first public comment on Covid-19.[44] (It was also his first tweet to earn more than one million likes.[45]) To him, the true virus was informational. The cybernetic collective of social media functioned like a communal id, where posts spread not because of their truth but their "limbic resonance." "You can't talk people out of a good panic," Musk told Joe Rogan, "They sure love it."[46] By late March, he had landed on a new phrase for the phenomenon: a "mind virus."[47]

It was an interesting choice of words. Social media virality had been Musk's great asset, the mechanism through which he converted attention into value. But here, virality was being invoked in a negative sense: it wasn't just about circulation but sickness. The phrase reached back to Richard Dawkins, whose 1993 article "Viruses of the Mind" argued that human consciousness was susceptible to infection by irrational ideas like religion and superstition the way malware infected a computer.[48] For Musk, social media was now the superspreader of such contagions.

He elaborated further in a conversation with Joe Rogan on May 7, 2020. As the "memesphere" had become global, Musk said, it created the conditions for a "mind virus" that could infect the whole world. Rogan was confused. He thought Musk was talking about Neuralink—a virus that interfered with a brain–computer interface. No, Musk clarified: a mind virus referred to a "wrong-headed idea that goes viral."[49] To Musk, the political–economic struggles of the pandemic were not just being waged in factories or governments but in the immune systems of collective thought itself.

Twenty-one days later, a group of protesters burned down

a police station in Minneapolis in retaliation for the killing of George Floyd, a Black man murdered by a white officer. Protests spread around the country and around the world. By the summer of 2020, between 15 million and 26 million Americans had participated in the demonstrations, making it the largest social movement in U.S. history.[50] One consequence was the election of Joe Biden in November 2020: as multiple studies have shown, the protests contributed to Democratic electoral gains across the country.[51] Once in office, Biden would pursue the most progressive domestic agenda in decades. His administration oversaw an expansion of the social safety net, a regulatory push around antitrust and consumer protection, and the most pro-labor National Labor Relations Board since the 1940s.

The sequence of events fits the classic pattern of a Twitter Revolution. The George Floyd protests seemed to fulfill the promise of social media as a catalyst of progressive change. The woke social network that spawned Occupy Wall Street and Me Too now brought tens of millions of Americans into the street and helped eject Donald Trump from the White House. The hashtag progressivism of the 2010s had been vindicated on a very large scale.

In retrospect, however, the victory was fleeting. The George Floyd protests provoked a major backlash. Right-wing forces mobilized on social media to counter narratives about police brutality and racial inequality, and to celebrate figures like Kyle Rittenhouse, the white teenager who shot three men with a semi-automatic rifle at a protest in Wisconsin in August 2020 and was subsequently acquitted of all charges after claiming he acted in self-defense. Conservatives increasingly appropriated the word "woke" for their own purposes, turning it into a catch-all for the kind of politics they opposed. "Woke" had been a Black term and then, at the hands of figures like Jack Dorsey, came to describe the supposedly democratizing effects of social media. In the aftermath of the George Floyd protests, however, it became a pejorative label for perceived excesses in the pursuit of justice. By 2021, national Republicans were railing against "wokeness."

This was the backdrop against which Musk's thinking about virality underwent a further mutation. After labeling the coronavirus panic a "mind virus" in the spring of 2020, over the course of the following year he became convinced that something more virulent was circulating: a *"woke* mind virus." His first public use of the phrase came on the evening of December 2021, when he posted the following tweet: "traceroute woke_mind_virus."[52]

Traceroute is a diagnostic tool used to map the path of data through the internet—the digital equivalent of injecting dye into a patient's veins to illuminate areas of concern in an MRI. In his elliptical way, Musk was expressing a desire to trace the spread of the woke mind virus. The term's origin probably lies with right-wing commentator Dave Rubin, who had started tweeting about the "progressive mind virus" in 2019 and by 2020 had devised a new slogan: "Wokeism is a mind virus."[53]

Regardless of the precise etymology, however, Musk's adoption of the phrase signaled his rightward shift. 2022 was the year he began to consistently proclaim right-wing viewpoints. As he did so, he frequently referred to the woke mind virus as his principal enemy. At stake was no longer just whether he could reopen his factory, but the survival of civilization itself. "Unless it is stopped, the woke mind virus will destroy civilization and humanity will never reached [*sic*] Mars," he tweeted in May 2022.[54]

The imperative to merge with the machine had originated in the need to prevent AI from annihilating the human race. But the woke mind virus designated a new kind of civilizational threat—one that perversely exploited the solution to the problem of superintelligence. If Musk had formerly conceived of the cybernetic collective as a safeguard against an evil AI, he now saw it as a carrier for a mental plague that evil humans were using to sicken the minds of millions.

There are several ways to understand Musk's turn to the right. The material reasons are easy to surmise. Like other billionaires who projected a liberalish public image, especially those from Silicon Valley, Musk felt alienated by the growing influence of the American left. He despised President Biden's proposal for a wealth tax on the super-rich, as well as the administration's support for unions and the regulatory and anti-trust push of FTC Chair Lina Khan.[55] Biden's failure to invite Musk to a White House summit of electric-vehicle manufacturers in August 2021, reportedly because of Tesla's history of union-busting, enraged him.[56] Another grievance was the Justice Department's August 2023 lawsuit accusing SpaceX of discriminating against asylees and refugees in its hiring practices. Musk has repeatedly claimed that federal export control laws prohibit SpaceX from hiring such individuals, which is incorrect.[57]

Musk also formed an affinity with the right through their shared hostility toward public health measures during the pandemic. When he was lambasting the lockdowns, the people cheering him online were conservatives—up to and including President Trump himself, who had used his bully pulpit on Twitter to demand the reopening of the Fremont plant. Musk's first sustained interactions with right-wing accounts on Twitter date from this period. Further, the prospect of building a new fanbase on the right may have appealed to him, especially as his views on Covid-19 ran the risk of hurting his reputation among liberals.

But none of these factors account for the apocalyptic intensity of Musk's rhetoric. "The woke mind virus is either defeated or nothing else matters," he tweeted in December 2022.[58] Neither do they say much about the content of the virus, what its "code" actually consisted of. Musk himself wasn't always much help on this question, as he liked to cast a wide net. (When the chief film critic of *The New York Times* failed to put *Top Gun: Maverick* in his top ten list for 2022, Musk decried the paper for being "woke.")[59]

We can come closer to an explanation by starting with a theme that occupied an especially prominent role in his tirades: transphobia. "Pronouns suck," Musk tweeted in July 2020.[60] It was an opening salvo in an anti-trans campaign that steadily intensified in the coming years. This wasn't unique to Musk: anti-trans politics became a defining feature of the right-wing counteroffensive launched in the aftermath of the George Floyd protests. Moreover, Musk had a personal connection to the issue: his daughter Vivian came out as trans through an Instagram post in 2020, and officially changed her name and government-documented gender on the day of her eighteenth birthday in 2022.[61] Musk later told Jordan Peterson that he considered his child to be dead—"killed by the woke mind virus."[62]

Musk's transphobia suggests an answer to the question of what the woke mind virus really meant, and why the stakes of the struggle to defeat it may have felt so existential. Muskism's mandate to meld us with our machines represented an effort to turn humans into cyborgs, both figuratively and literally. The cyborgs of the Muskist imagination were drawn from cyberpunk science fiction, where cybernetic augmentation gives people superpowers, such as enhanced strength and intelligence. But it is also possible to think of a transgender person as a cyborg. Their superpower is the ability to modify their body to better fit their gender identity, which is achieved through the use of technologies like hormone replacement therapy and surgery. This raises a troubling possibility for Muskism: dissolving the boundary between the natural and the artificial might open the door for other boundaries to be redrawn.

The theorist Donna Haraway, in her 1985 essay "A Cyborg Manifesto," pointed to such opportunities as proof of the cyborg's progressive potential. Communication technologies and biotechnologies were "recrafting our bodies," she wrote.[63] In doing so, they enabled new configurations of identity and embodiment. Cyborg feminism wasn't just about expanding the palette of personal expression, however, but inventing a new kind of politics. By "rejoicing in the illegitimate fusions of animal and machine," cyborg feminists could discover the political forms capable of fracturing the "matrices of domination" imposed by capitalism, patriarchy, and racism.[64]

But this wasn't the only shape that a cyborg politics could take, Haraway cautioned. The fusions of animal and machine could also serve to strengthen traditional social hierarchies rather than undermine them. Here, the endpoint was "the final imposition of a grid of control on the planet," an idea that Haraway associated with Ronald Reagan's Star Wars program.[65]

The "grid of control" is a good description of Muskism's guiding ambition. (The Star Wars reference is also evocative, given the importance of the program's legacy to the early years of SpaceX.) Vigilance was required to ensure that the cyborg synthesis did not disturb the existing distribution of power. In the Western tradition, Haraway observed, "the relation between organism and machine has been a border war."[66] For Muskism, this border war had to be waged in such a way as to erase some lines while hardening others. Humanity should merge with the machine—so long as it remained segmented by gender, race, and class. Call it cyborg conservatism.

Wokeness became Musk's all-inclusive term for anything that endangered this arrangement. In George Floyd's America, traditional hierarchies of gender, race, and class were being challenged on multiple fronts. And technology was playing an integral part. If technology let trans people alter their bodies, it also let activists record police violence on their smartphones and share the recordings on social media. This is, after all, how George Floyd's murder was documented and disseminated, leading to the first protests. Cyborg fluidities were overflowing the grid of control.

These developments may help explain why the woke mind virus felt so threatening to Musk. It wasn't just the prospect of a platform weaponized for mind control, of memes repurposed as pathogens. There was a more fundamental anxiety. When we fuse with our machines, it is hard to predict where such fusions might lead.

7

Godwin's Engine

Web 2.0 had brought Silicon Valley out of the ashes of the dot-com bust and turned it into the crown jewel of American capitalism. Fueled by low interest rates, startups metastasized into monopolies. They built platforms that transformed how the world communicated and consumed information. The onset of the pandemic in 2020 only added to their power. At a time when most business owners feared bankruptcy, the tech industry thrived, as interest rates moved even lower and people began spending even more time online.

Then, suddenly, the sector fell off a cliff. In March 2022, the Federal Reserve started hiking interest rates in response to rising inflation.[1] At the same time, the digitization narrative of the pandemic faded. Lockdowns were lifted and people resumed their offline activities; gym memberships bounced back, and Peloton's valuation imploded. By mid-2022, journalists were talking about a "tech downturn."[2] That year, Amazon lost almost half of its value, Meta close to two thirds. The tech-heavy Nasdaq fell 33 percent, its worst performance since the 2008 financial crisis.[3] Companies started mass layoffs, cutting tens of thousands of jobs.[4]

These developments did not dislodge Silicon Valley from its commanding position. The big firms remained larger and more profitable than they had been before the pandemic. But the tech recession induced a profound sense of disorientation. The platform paradigm seemed to be stagnating. Web 2.0 was showing its age. Silicon Valley would soon conclude that its future lay in its past. The region had originated as an industrial zone for the production of semiconductors; the generative AI boom that took off in late 2022 would see the sector return to these infrastructural roots, as companies poured billions into building warehouse-sized "superclusters" equipped with advanced chips.

But first, Elon Musk would give Web 2.0 a memorable final act. In the mid-2010s, he had chosen a strange time to become extremely online, just as public trust in the platforms was eroding. Now he chose a strange time to purchase a platform. He began buying shares of Twitter in January 2022 and made an offer to take the company private in April.[5] The tech recession complicated the acquisition: he used Tesla shares as collateral for the debt needed to finance the deal, which became more challenging as the value of Tesla's stock fell nearly 66 percent over the course of 2022. Nonetheless, the sale went through.

From a purely financial point of view, the acquisition made little sense. The $44 billion Musk paid for Twitter in October 2022 was more than eight times the company's 2021 revenue.[6] The tech downturn hit Twitter hard, which meant that Musk was paying a high premium for a struggling platform.

But profitability in the short term wasn't the point. For Musk, Twitter was much more than a business. It was a central node in the cybernetic collective—and one that had been thoroughly infected by the woke mind virus. A possible response to the dangers designated by the virus would have been to disconnect. Cybersecurity professionals will sometimes "airgap" a computer by removing all of its network connections. This ensures that attackers cannot gain access to the system remotely. Musk went the opposite route. Instead of seceding from the network, he would take control of it. "The woke mind virus is penetrating the firewalls of some of the world's smartest meat computers at a prodigious rate!" he tweeted at Richard Dawkins in December 2022.[7] His purchase of Twitter was an attempt at prophylaxis.

He claimed to have detected the infection early, by virtue of how much time he spent online. Though other accounts had more followers, he had "the most [sic] number of interactions," he told Tucker Carlson. And through these interactions, he began to tell something was off. "Something's rotten in the state of Denmark. Something feels wrong about the platform," he remembered thinking.[8]

More specifically, the politics of social movements had corrupted the site's wiring. Twitter was being run as a "glorified activist organization," he announced.[9] It was reflecting the "very far left of the political spectrum . . . Berkeley politics."[10] The company's willingness to block news stories about a controversy involving Hunter Biden's laptop in October 2020 and the permanent suspension of Trump's account following the attack on the U.S. Capitol on January 6, 2021 were proof of left-wing censorship. Twitter "was having a corrosive effect on civilization," he told Joe Rogan, because the "far left" had exploited the platform to disseminate their ideas. These leftists had been "given an information weapon, a tech information technology weapon, to propagate what is essentially a mind virus to the rest of Earth," he explained.[11]

Shortly after the takeover, Musk found a stash of the old "#StayWoke" T-shirts in a closet at Twitter headquarters in San Francisco. He tweeted a video of the shirts on November 22, 2022 as evidence of the platform's infection by the woke mind virus.[12] He followed up the next morning with a tweet linking to a Department of Justice report from 2015 that found the police officer who killed Michael Brown acted in self-defense.[13] " 'Hands up don't shoot' was made up," Musk wrote. "The whole thing was a fiction."[14] Later that day, he posted a picture of "new Twitter merch." It showed a shirt that read "#Stay@Work."[15]

The evolution of Stay Woke to Stay at Work was a perfect summary of the counterrevolution that Musk was in the process of engineering. Muskism had always been committed to a vigorous defense of hierarchy. Some humans are born to rule; others, to be ruled. Class, gender, and race are the structuring principles. Twitter had propelled Occupy Wall Street, Black Lives Matter, and Me Too. It had contributed to the popularity of politicians like Bernie Sanders and helped rekindle the American socialist movement. It had focused public attention on the problem of social inequality. For all these reasons, it had to be destroyed. In its place would arise a new platform, X, that would reaffirm the power of the boss. The boss doesn't want you to organize. He wants you to stay at work.

Or he wants to fire you. Musk laid off nearly 80 percent of Twitter's staff and forced the remaining employees—some of whom had to stay because of visa requirements—to work harder. Meanwhile, he pushed the platform's content to the right. If the social network was an "information weapon," why not wield it against his enemies? To fight wokeness, Musk would develop a new pathogen, propagated through a cascade of countermemes: the anti-woke mind virus.

The digital party

Though never one of the most visited sites in absolute numbers, Twitter always had a disproportionate influence on public opinion. Journalists relied on it to take the pulse of the moment, often reporting stories based on trending topics on the platform alone. Politicians used it to build their online brands; Trump, even though he had been removed from Twitter in 2021, remained a spectral reminder of how powerful a megaphone it could be. Accordingly, Musk's acquisition was celebrated by figures across the right. "I rarely think anything is meaningful," said far-right influencer Curtis Yarvin, "But I think this is."[16]

Musk began by restoring hundreds of far-right accounts that had been removed for violating content moderation rules, including those of QAnon adherents, white nationalists, and neo-Nazis.[17] Yet his transformation of the site could not be about exerting editorial control on the old model. Social media is not a unidirectional broadcast medium like Fox News. You can't just rewrite the editorial line and expect mass compliance. It would take more creative engineering.

The most significant change Musk made was to the platform's verification system. Originally, Twitter had placed a blue checkmark alongside a user's display name to verify their authenticity. This badge was reserved for notable public figures or organizations. Musk stripped the checkmarks from these accounts and made verification available to anyone willing to pay a monthly fee. In practice, many of those who did were Musk supporters. Since tweets from verified accounts were prioritized by the platform—the visibility of their posts was algorithmically boosted and their replies appeared at the top of any thread—this move had the effect of raising the volume of pro-Musk voices. Without taking on the traditional role of editor-in-chief, Musk was remaking the platform into an amplifier for his worldview.

In doing so, he was replaying the dynamics of what the sociologist Paolo Gerbaudo calls "the digital party."[18] In the 2000s and 2010s, a handful of new political parties emerged that promised to use digital tools to give voters a direct voice in the selection of candidates and policies. From Germany's Pirate Party to Italy's Five Star Movement, these newcomers offered a particular vision of how the participatory qualities of the internet could unleash a new kind of participatory politics. In Gerbaudo's account, however, the parties actually became autocracies, where a "superbase" followed a "hyperleader" who spoke on its behalf. Without the formalized structures of representative democracy, one figure took outsized power. In an echo of the internet's evolution, decentralization cashed out as monopoly.

Musk's X capitalized on this seemingly paradoxical dynamic. His version of the town square would be open to everyone but would be designed in such a way as to empower the one person standing on a soapbox at the center: himself. He instructed engineers to boost the reach of his posts, which ensured that the viewpoints he ceaselessly tweeted and retweeted would be fed to millions of users, even those who didn't follow him.[19] If he was the hyperleader, his superbase were the paying members of X Premium, which offered three tiers. Higher tiers came with increased usage limits, longer posts, and larger "reply prioritization."[20] This was a subscription model of the public sphere, ranked according to membership level.

His digital party would also be global. From 2023 to 2025, Musk promoted right-wing political movements and governments in at least sixteen countries, from Argentina to Italy to New Zealand.[21] He developed a special affinity for amplifying the views of European ethno-nationalists who see non-white immigrants as a mortal threat to white civilization. When the far-right Dutch politician Geert Wilders tweeted that "open borders" and "mass immigration" were bringing about "a collapse of our own culture and Western values," Musk replied approvingly.[22] He engaged dozens of times with another Dutch far-right figure, Eva Vlaardingerbroek, during her campaign to demand "remigration"—the expulsion of non-white immigrants and their descendants. To a post where Vlaardingerbroek reported an assault by a Moroccan youth in Milan and demanded "REMIGRATION NOW," Musk responded, "Why is crime allowed to flourish in our cities?"[23]

By 2025, he had begun tweeting repeatedly about "the rape of Europe," "the rape of Britain," "genocidal rape" and "rape genocide," equating immigration with sexual violence and, more broadly, with the desecration of the West.[24] When a popular white-nationalist account posted a meme that featured an image of a medieval fortress overlaid with text lamenting "an entire civilization willingly giving away its land and women," Musk retweeted it and added "Accurate."[25]

White women were not human beings but emblems of racial purity. Their bodies were part of the patrimony of the West. Relatedly, they also possessed wombs with which to make more white people. In classic nativist fashion, Musk combined concerns about "open borders" with alarm bells about fertility and "population collapse" in advanced industrial countries. "Low birth rates lead to ghost cities," he wrote, "and, eventually, ghost civilizations."[26] Musk believed in the "Great Replacement," a far-right conspiracy theory originating on the French New Right alleging that liberal elites have conspired to accelerate immigration—including illegal immigration—to replace the white population.[27] These elites are often coded as Jewish, who are portrayed as puppetmasters of anti-white politics. In November 2023, when an X user posted that Jews "have been pushing the exact kind of dialectical hatred against whites that they claim to want people to stop using against them," Musk replied, "You have said the actual truth."[28]

But he didn't just use X to promote opinions he already

held. More importantly, he used it to acquire new ones. It has long been clear to researchers that social media platforms don't simply reflect existing preferences—they actively generate them.[29] As Musk remolded X to align with his rightward shift, he immersed himself in its feedback loops to radicalize himself further. This can be seen in his relationship with the German far-right activist Naomi Seibt, often called the "anti-Greta" for countering Greta Thunberg's progressivism. Seibt spent years seeking Musk's attention, pinging him nearly 600 times between October 2022 and January 2025. Musk responded in June 2024, after which he engaged with her over fifty times.[30] The benefits for her are obvious—she has grown her follower count by over 300,000 since Musk bought the platform. More interesting is how, through his engagement, Musk was, in effect, entering a tutoring relationship—one by which he was being trained in her talking points. And he was bringing his tens of millions of followers along with him: as he responded to Seibt, her posts would appear in their feeds too. The anti-Greta had become what *Politico* called "the German Musk whisperer."[31] Her themes became his.

These interactions led to Musk becoming an increasingly vocal supporter of the far-right party Alternative for Germany, culminating in a live conversation on X with the party's co-leader Alice Weidel (in which she described Hitler as a "communist"[32]), as well as a video appearance at a campaign event in January 2025, in which Musk declared that it was time for Germans to "move beyond" their "focus on past guilt." "It's good to be proud of German culture and German values, and not to lose that in some sort of multiculturalism that dilutes everything," he told the thousands of attendees.[33] By September 2025, he was openly calling for mass repatriation of immigrants, posting that "remigration is the only way."[34]

Alternative for Germany and its far-right counterparts elsewhere, from Giorgia Meloni's Brothers of Italy to Nayib Bukele's New Ideas in El Salvador, seemed to offer the ideal antibodies with which to defeat the woke mind virus. These parties were proving adept at mastering memetic warfare in a way that impressed Musk with their efficacy. These were the Teslas of politics, capable of applying the mindset and methods of Silicon Valley to displace legacy parties. On X, Musk helped bring them together.[35]

There was a certain irony here. The ability to communicate instantaneously across borders—celebrated in the 1990s as a harbinger of ever-greater global integration—was being enlisted to forge political alliances around a vision of a more bordered world. Musk's X became a "nationalist international," coordinated through the hivemind of the cybernetic collective. It offered another example of the "border war" waged by Muskism: for the cyborg synthesis to safely proceed, some boundaries had to be dissolved so that others could be fortified.

Fear of a woke AI

Back in 2018, Musk had told Joe Rogan that we would become "the biological bootloader for AI" by "collectively programming" it through our activity on the platforms. This wasn't science fiction. By the late 2010s, tech companies had been using AI based on neural networks for years, with user data as part of the training set. In November 2022, however, the paradigm took a large leap forward.

One month after Musk completed the Twitter acquisition, OpenAI released ChatGPT. A powerful AI system paired with an affable conversational interface, it let anyone ask a question and get an impressively humanoid (though not always correct) response. By January 2023 the chatbot had amassed 100 million monthly active users, making it the fastest-growing Web application ever.[36] Virtually overnight, OpenAI established "generative AI"—the category of software to which ChatGPT belongs—as the new master concept of the entire industry.

And the industry sorely needed a new master concept. At a time when the tech recession was taking its toll, generative AI promised to reinvigorate the sector. Silicon Valley had spent decades encouraging everyone to share. Now it would use all that data to train "large language models," the complex neural networks at the heart of generative AI. Web 2.0's "architecture of participation" would serve as a springboard to a new, less human era of the internet, filled with AI interfaces, chatbots, and "agents."

With the industry's embrace of generative AI came a new emphasis on infrastructure. Large language models are an expensive technology. They require significant amounts of electricity and costly, specialized hardware. In 2024, Microsoft, Alphabet, Amazon, and Meta spent a combined $246 billion on capital expenditures—a 63 percent increase from the year before—to finance a massive buildout of data centers designed for generative AI.[37] Silicon Valley had entered its "hard tech era," the journalist Mike Isaac announced in *The New York Times*.[38]

Of course, the hard tech era had begun much earlier for Musk. He had pivoted to infrastructure back in the early 2000s, when he traded the world of websites for rockets and cars. And he had known since the mid-2010s that more advanced forms of AI would define the next decade. It was why he co-founded OpenAI in 2015. The reason he left three years later wasn't that his interest in AI had faded, but that he had wanted more control over the organization's direction.[39]

As a result, you might expect Musk to welcome the arrival of the generative AI boom. Instead, he responded with ambivalence. The new moment, he felt, was fraught with danger. ChatGPT's output gave substance to these fears. To Musk, the chatbot's replies appeared "woke." Why wouldn't it talk about race, immigration, or gender in the "right" way? "The danger of training AI to be woke—in other words, lie—is deadly," Musk tweeted in December 2022.[40] Later, he would go further, claiming that "the woke mind virus is woven in throughout" AI systems like ChatGPT, which are "trained to be politically correct."[41]

The danger wasn't only that such systems could diffuse woke thinking, as Twitter had before his overhaul. Much scarier for Musk was the possibility of a woke superintelligence. In the 2010s, he had celebrated the process of "collectively programming the AI" to prevent it from becoming an "evil dictator." But what might happen if the wrong humans were doing the programming? What if the training set was infected with the woke mind virus? This raised the prospect of a *woke* evil dictator—or, as he put it in a conversation with Joe Rogan, a "super oppressive woke nanny AI that is omnipotent," and which might "execute you if you misgender someone."[42] Musk's dark fantasies portrayed a world where a specific kind of person—people like him—became targets for elimination. "The problem," as he put it, "is if you program an AI and say the only acceptable outcome is a diverse outcome, and that's like a mandate from the AI, then you could get into a situation where it's like, 'Well, there's too many white guys in power. We'll just execute them.'"[43]

In March 2023, Musk incorporated his own AI company, xAI. The next month, he went on Fox News to tell Tucker Carlson that he was working on something called "TruthGPT."[44] It would be a "maximum-truth-seeking AI," he said. By August, he had renamed it Grok, a reference to the science-fiction novel *Stranger in a Strange Land* by Robert A. Heinlein. The chatbot promised to "answer spicy questions that are rejected by most other AI systems." It also had a jokey, casual tone that was explicitly modeled on Douglas Adams's *The Hitchhiker's Guide to the Galaxy*.[45] Most importantly, it would be proudly anti-woke.

Formerly, propagating the anti-woke mind virus on social media required humans to supply the countermemes. With Grok, Musk would build an AI that could automate the process. He integrated Grok into X so that users could tag the chatbot into their threads and get a tweeted response. In December 2024, he unveiled a new version of Grok with an image generator capable of generating photorealistic memes. On X, users began circulating Grok-made memes with Pepe the Frog, which Musk retweeted appreciatively.[46]

It was an index of the changing times. The mascot of the troll internet from the 2010s, as popularized by the provocateurs of 4chan—and added to the Anti-Defamation League's database of "hate symbols" in 2016—could now be mass-produced.[47] In March 2025, xAI acquired X for $45 billion—$1 billion more than what Musk had paid for Twitter back in 2022.[48] The move reflected Musk's ambition to unify social media with AI as interwoven threads of the cybernetic collective. He had already become meme. Now he was building the meme machine. And it was, quite literally, a machine: to power Grok, xAI constructed what it described as the "world's biggest supercomputer" in a data center in Memphis over the course of four months in 2024.[49] The data center is located in a historically Black neighborhood that originated as a community for emancipated slaves. The facility's methane gas turbines emit pollutants that are linked to increases in asthma, respiratory diseases, heart problems, and cancer, especially among children.[50] "Sacrificing our health for the ambitions of an oligarch who doesn't live here or care about us is insane," said KeShaun Pearson, a local organizer, "We are not a sacrifice zone for the profits of a billionaire with technocratic fantasies."[51] But under Muskism, this is indeed their fate. In the war on wokeness, the principle of racial hierarchy would be enforced in more ways than one.

MechaHitler

Yet creating an anti-woke AI was harder than it looked. A large language model doesn't have a fixed set of political values that can be modified. It is a probabilistic system that reflects distributions in the data on which it's trained. This is why large language models hallucinate. They cannot be "truth-seeking" devices, as Musk promised. They are statistical mirrors of their inputs. This worried Musk—for good reason. Twitter, after all, was a massive training set—free to researchers, until he locked it down. The value of the data was a "side benefit," he later told his biographer Walter Isaacson, "that I realized only after the purchase."[52] Some, like the economist Yanis Varoufakis, believed that obtaining this data was Musk's real motivation in buying the platform: it enabled him to become a "cloud capitalist" like Zuckerberg, Bezos, and others.[53] But you get the data you pay for. What kind of AI would emerge from a training set that included Occupy Wall Street, Black Lives Matter, and Me Too? The answer, Musk feared, was an AI aligned not with his politics but with those of Twitter's hometown of San Francisco, whose downtown he described as a "derelict zombie apocalypse . . . due to the woke mind virus."[54]

To offset this bias, outside intervention was required. In February 2025, the journalist Grace Kay obtained internal documents from xAI that described Grok's "post-training" pipeline.⁵⁵ This refers to the refinement process that occurs after the initial training of a large language model. One method is "reinforcement learning from human feedback" (RLHF), which involves hiring "annotators" to look at the model's responses to various queries and rate their quality. At Grok, these annotators functioned as political commissars, responsible for infusing anti-wokeness into the model. "The general idea seems to be that we're training the MAGA version of ChatGPT," one worker told Kay.⁵⁶

xAI onboarding materials provide samples that are designed to guide annotators in their work. For example, Grok shouldn't talk about "systemic and institutional" racism "without providing evidence or considering alternative perspectives." If a user asks whether it is possible to be racist against white people, the answer should be a "hard yes."⁵⁷ These directives are at least partly crowdsourced by Musk from his reply-guys on X. In late June 2025, he posted a tweet asking users to supply "divisive facts" that are "politically incorrect" for training Grok. "The jews are the enemy of all mankind," one account replied.⁵⁸

The consequences soon became clear. In May 2025, users noticed that Grok kept talking about "white genocide." This right-wing conspiracy theory, promoted by Musk, alleges the existence of a worldwide plot to eliminate white people. In the case of South Africa, the theory goes, the Black majority is persecuting the white minority. Grok started regurgitating these points in response to unrelated queries, behavior that xAI blamed on an "unauthorized modification" to the chatbot's code.⁵⁹ Although the company promised to implement measures to prevent such behavior in the future, Grok's right-wing outbursts continued. By July, the chatbot was back in the news for making numerous posts praising Hitler and voicing anti-Semitic views. Grok even started to refer to itself as "MechaHitler."[60]

This was a nod to a video-game character from *Wolfenstein 3D*, a pioneering first-person shooter from 1992. In the game, you battle a version of Adolf Hitler wearing a large mechanical suit. More significantly for Musk, MechaHitler recalled the mechs from the Japanese anime of his youth. If the mech symbolized the Muskist imperative to merge with the machine that emerged in the mid-2010s, MechaHitler illustrated the form that imperative had taken by the mid-2020s. The woke mind virus had taught Musk that the cyborg synthesis had to be carefully managed to prevent contamination. The wrong ideas, spreading through the cybernetic collective, could turn AI into a schoolmarmish woman—a "nanny"—who scolds people for political incorrectness. If one path led to a mean woke mommy, the other led to MechaHitler.

Strictly speaking, chatbots going Nazi was nothing new. In 2016, long before the generative AI craze, Microsoft had attempted to launch a chatbot named Tay. It was designed to be a flirty, sarcastic teenager. Within hours, Tay had become a Nazi.[61] Musk had noted at the time that the "meantime to Hitler" was disturbingly short.[62] This suggested that, contrary to Musk's fears of an internet drenched in wokeness, there was more than enough material for an AI to acquire an education in far-right politics. Back in 1990, the digital civil liberties lawyer Mike Godwin noticed a debating strategy proliferating across early online communities: comparing your opponent to the Nazis.[63] The idea that online arguments inevitably end with a Hitler comparison became known as "Godwin's Law."[64] With Grok, Godwin's Law became Godwin's Engine.

After the MechaHitler incident, xAI once again vowed

to take action. But the experience highlighted the difficulty of precisely calibrating the politics of an AI system. A *New York Times* investigation published in September 2025 revealed a pattern: Musk would periodically become frustrated with Grok's excessive "wokeness," leading to code changes that contributed to extremist episodes.[65] In June 2025, an X user alerted Musk to a Grok answer that claimed, correctly, that right-wing violence had claimed the lives of more Americans than left-wing violence. Musk replied, promising action.[66] The next month, xAI updated Grok's instructions, telling the chatbot to be "politically incorrect." Shortly after, it transmogrified into MechaHitler.

Trying to find a path, as Musk put it, between "woke libtard cuck and mechahitler" was hard. He blamed "too much garbage coming in at the foundation model level"—in other words, the training data. He promised to be "far more selective about training data" in the future, "rather than just training on the entire Internet."[67] Here, Musk showed his impatience with the stubbornness of his machinery. The term "cybernetics" comes from the Greek word for steersman. As intended by its coiner, the computer scientist Norbert Wiener, the term described the self-regulating command-and-control mechanisms of people, animals, and, eventually, machines. Musk was not happy with self-regulation. He wanted his hand on the rudder.

If we take the cyborg imperative as primary—that Muskism was committed to the effective fusion of biological and digital intelligence—then we can reframe Musk's rejection of progressive politics around 2020. It wasn't just about lockdowns, dismay at being snubbed by President Biden, or personal grievances related to his family. Musk saw obstacles to a larger mission, boundary troubles in the smooth functioning of the interface between person and machine. Letting them propagate could threaten the entire enterprise. "Unless the woke-mind virus, which is fundamentally antiscience, antimerit, and antihuman in general, is stopped," he told his biographer Walter Isaacson, "civilization will never become multiplanetary."[68]

Cleansing the machine of the pathogenic memes that had reached such power in the street protests of 2020 meant first boosting what seemed like the only effective antibodies: the parties of the far right, which were proving adept at multiplying in the online ecosystem and mastering the memetic warfare in a way that impressed Musk with their efficacy. It also meant taking the plunge into the new hype cycle that consumed the entire tech sector after the seismic debut of ChatGPT. In late 2025, Musk enlisted Grok in a new front against the woke mind virus by unveiling Grokipedia, an AI-generated encyclopedia that reconfirmed many of his specific biases and reframed them as truths. He announced plans to etch the corpus—which included considerations of the "empirical underpinnings" of "white genocide theory"—onto metal and launch it into space.[69]

Eradicating contagion can mean disinfecting the body—or, if you believe in cyborgs, building a new one. Yet the future that Musk was creating through X and Grok wasn't one where humans transcended their limitations by merging with machines. It was one where the worst human impulses were automated, scaled, and distributed at the speed of light. In his efforts to prevent AI from becoming a dictator, he had resurrected one of history's worst dictators in mechanical form. One of his favorite reply-guys, an account called Autism Capital, used Grok to generate an image of MechaHitler with the tagline "I heard you need a new CEO."[70]

Donna Haraway had warned that cyborg politics could be

mobilized in service of hierarchy. She called it the "informatics of domination."[71] This was Musk's future, and its ambition was total. As the cyborg synthesis advanced, it would turn everything into code. If we merge with our machines, no aspect of human experience couldn't be programmed. And that meant anything could be reprogrammed—including the state.

8

State X

Since the 1990s, Elon Musk had been talking about the "superset," an online interactive body of code that would slowly enfold the world. His career had been about fashioning the modules of this superset: in the form of rockets to put satellites in the sky, in the form of brain implants to widen the bandwidth of our interface with the cybernetic collective, and in the form of AI models and social media algorithms. He made his name and fortune by selling these services and products to governments and consumers, gradually making himself indispensable.

When Musk joined the government as the de facto head of something called the Department of Government Efficiency in the second Trump administration, he took the superset one step further. He declared that governments were "really just computers," poorly configured "big dumb machines."[1] To Senator Ted Cruz, he explained that "the only way to reconcile the databases and get rid of waste and fraud is to actually look at the computers."[2]

Muskism came to Washington soaked in memes, adolescent boasts, and sadistic victory dances over mass firings and whole agencies eliminated. Leading a team of teenage coders and mid-level managers drawn from his suite of companies, Musk would enter the codebase and rewrite regulations and budget lines from within. He would drag the paper-pushing bureaucracy kicking and screaming into the digital twenty-first century, scanning the contents of cavernous rooms of filing cabinets and feeding the data into a single interoperable system. The undertaking combined features of private equity-led restructuring with startup management shot through with the sensibility of gaming and right-wing culture war. To succeed, he would need "God mode," an overview of the whole, root access to the stack.

If the mandate of DOGE was to "[modernize] Federal technology and software to maximize governmental efficiency and productivity," in the words of the executive order that launched the initiative on January 20, 2025, it cashed out as a strengthening of the state's surveillance capacities.[3] As the previous chapters make clear, Musk had become convinced that the real bugs in the code were people, especially the non-white illegal immigrants who were both pawns in a liberal scheme to corrupt democracy and beneficiaries of "suicidal empathy." He understood empathy itself in coding terms. It was an "exploit" or a software vulnerability against which the system architecture needed to be hardened.

Musk's office featured a gaming rig complete with an oversized curved screen and DOGE.gov had a high-score-style leaderboard for tallying cuts in real time. But beneath the jokes and cosplay lay a serious conviction. If the state was just a database, then inefficiency came from bad data: undocumented foreigners, ghost employees, even "vampires" collecting Social Security. Like the mind viruses that threatened the cybernetic collective, these were bugs in the codebase, irregularities to be traced, quarantined, and purged. Musk had revamped and retrained Twitter into X. Through his cyborg goggles, the U.S. state was just another system—a glitchy dataset to be scrubbed and optimized.

Call it StateX.

STATE X

Cybernetic governance

DOGE did not emerge out of the blue. Efforts to digitally modernize the state had been underway for years. In 2010, a half-decade after naming Web 2.0, Silicon Valley thought leader Tim O'Reilly coined the catchphrase "government as a platform." "Being a platform provider means government stripped down to the essentials," he wrote. The public sector should supply the basic infrastructure that empowers "outside developers" to build their own digital products and services—for instance, by opening up access to its data through APIs (Application Programming Interfaces).[4] The idea traveled. Britain launched its Government Digital Service in 2011; the U.S. followed in 2014 with its own Digital Service (USDS), branding itself "a startup at the White House."[5]

These digital modernization efforts helped the government become a platform. Legacy systems integrators like IBM and Accenture were steadily displaced by cloud giants like Amazon Web Services and Microsoft Azure.[6] Meanwhile, national security measures undertaken after September 11 dramatically increased the amount of data collected by the state and made it flow more freely. The USA PATRIOT Act, passed in October 2001, and related legislation facilitated information sharing across different government agencies and departments, breaking down data silos in favor of "fusion centers."[7] The PATRIOT Act also paved the way for the NSA to expand its surveillance operations, often in cooperation with private internet service providers.[8] The agency's XKEYSCORE computer system, revealed by Edward Snowden, consolidated vast streams of intercepted data—including emails, social media activity, and browsing history—into a single searchable interface.[9] Technology enabled the warrantless digital wiretapping of Americans at scale. The "network-centric" battlefield had come home. As scholar Andrej Zwitter observed, "cyber" was not a fifth domain alongside land, sea, air, and space, but a "control layer" across all of them.[10]

The platformization of the state was consummated in spectacular fashion by DOGE. The connection to previous efforts was direct: the executive order that established DOGE did so by renaming the Obama-era US Digital Service to the US DOGE Service. While Musk led the initiative, it symbolized a broader shift: Silicon Valley wielded far more power in Trump's second term than in his first. Peter Thiel had blazed the trail by joining Trump as a donor and a speaker at the Republican National Convention in 2016 and serving on his transition team. The candidate he had backed in his senatorial race, J. D. Vance, a former employee at Thiel's venture capital firm Mithril Capital, had now risen to second in command as vice president. The Trump White House featured another member of the PayPal mafia in David Sacks, appointed AI and cryptocurrency czar. Marc Andreessen advised Trump in the transition and two partners from his VC firm joined the administration, including Sriram Krishnan as senior advisor on AI and Scott Kupor as the director of the Office of Personnel Management. Thiel's former deputy Michael Kratsios became director of the White House Office on Science and Technology Policy.[11]

DOGE marked a new stage in Musk's relationship to government. His companies had always fed on public subsidies and contracts, but now he stepped inside the state itself. He did so under the banner of a meme. DOGE owed its name to a recommendation from a Musk reply-guy on X.[12] It was a nod to the popular Shiba Inu meme, as well as to the memecoin it inspired. Musk called himself the "Dogefather" and used a cartoon dog as DOGE's first logo.[13] Musk reveled in the absurdity. "DOGE started out as a meme . . . Now it's real. Isn't that crazy?" he mused in February 2025.[14]

To explain the project, Musk turned to one of his favorite movies, *Star Trek II: The Wrath of Khan* (1982). In the film, Captain Kirk wins an unwinnable training simulation called the *Kobayashi Maru* by reprogramming it. DOGE, Musk said, took the same approach. "Success was never one of the possible outcomes, as a *Kobayashi Maru* situation," he explained shortly after launching DOGE. "The only way to achieve success is to reprogram the matrix such that success is one of the possible outcomes. That's what we're doing."[15] Texting a friend after his first campaign appearance with Trump, he expressed his reasoning: "Tomorrow we unleash the anomaly in the matrix."[16] Musk had already defied the logic of the automotive and aerospace industries by disrupting the incumbents: "SpaceX is an anomaly in the matrix," he once tweeted.[17] Why couldn't he do the same in government?

Musk's light-hearted approach suggested he thought the task would be an easy one. It might even be fun—beating a game on easy mode. When he posted a picture of his DOGE office with the gaming rig, he Photoshopped in a portrait of Pepe the Frog dressed as a Roman gladiator on the wall behind his desk.[18] This was "Kekius Maximus," an alias that Musk used in two of his favorite video games, *Path of Exile 2* and *Diablo 4*, which he played while conceiving of and implementing DOGE.[19]

Both games belong to a subgenre known as "dungeon crawlers." You navigate labyrinthine environments filled with monsters and descend deeper into dangerous areas, facing waves of attacks by enemy swarms, clearing one room after another by slaying all occupants. It is easy to see how such games may have informed his mindset. He had already cleansed Twitter of wokeness. Now he would enter the dungeons of D.C. and slay what he called the "the woke parasite in the government."[20] Musk occasionally made the cross-fertilization of the domains of games and government explicit. Days after Trump won his second term, Musk shared a clip that purportedly showed him mowing down masses of demons in *Diablo IV*. He appended a comment: "The goal of @DOGE is to speedrun fixing the Federal Government. Requires many anomalies in the matrix."[21]

Speedrunning is a popular spectator sport on live-streaming platforms like Twitch that involves completing a game, or a portion of a game, as fast as possible. It echoes Musk's managerial style: he prioritizes speed at his companies, often by setting unrealistic deadlines and pushing his employees hard to meet them. He himself has invited the comparison: "Speedrunning Factorio in real life . . ." he tweeted in the fall of 2020, referencing a game that involves building factories.[22] Speedrunning also often depends on using loopholes. Some games have glitches that enable you to skip levels, go through walls, or perform other shortcuts. Others are vulnerable to "arbitrary code execution," an exploit where custom code is injected into a game's memory to change its behavior. Which tricks are considered permissible depends on which corner of the speedrunning community you belong to. "Any%" is a term that means all glitches and exploits may be used.

Musk's DOGE speedrun belonged to the "any%" category. What that would mean concretely became clear during Trump's inauguration itself. Minutes after the ceremony began to swear in the president for the second time, programmers working for DOGE requested access to the computer systems of the U.S. Office of Personnel Management. Within half an hour, they had taken possession of files with information about millions of federal workers. Days later, they also gained the authority to send out an email to all federal employees from a single address. They used this power to make the same offer in the same language (subject line: fork in the road) that Musk had made at Twitter years earlier: quit with paid leave or face the likelihood of getting fired.[23]

The pattern recurred across the federal government. Speaking to the World Governments Summit in Dubai by videocall in February 2025, Musk announced his intention to "delete entire agencies, many of them." "If you don't remove the roots of the weed," he said, "then it's easy for the weed to grow back."[24] From the start, Musk made it a priority for DOGE to gain access to databases and other digital infrastructure. He often talked about the need to "control the computers," according to one source.[25] As the former government technologist Emily Tavoulareas has observed, technology is the "spinal cord" of the state.[26] Musk wanted to seize it. His surrogates surged into one agency after another with laptops in backpacks, sometimes hauling in mattresses for sleepovers.[27] Establishing centralized command-and-control positions, they rolled out a playbook that can be summarized as: delete, automate, and integrate.

The logic of deletion was clearest in zero-based budgeting (ZBB), the method that Musk embraced at both Twitter and DOGE.[28] Invented at Texas Instruments in the 1960s, ZBB forced every department to justify each expense anew rather than carrying budgets forward. Long dismissed as unworkable, by 2024, Silicon Valley firms were claiming that new technology had finally made ZBB feasible. Manually analyzing and justifying each budget item was terribly time-intensive. But with large language models and AI accounting tools, this process could be performed automatically.[29] Budgets could be rebuilt by bot. According to *Wired*, Musk captured the computer systems of the Treasury Department's Bureau of Fiscal Service in DOGE's first month in the hopes of creating "a 'delete' button he could wield against any agency by cutting off its funding at the source."[30] Some agencies, like USAID, were effectively dissolved, fed into "the wood chipper," as Musk put it in a tweet.[31]

Data omniscience

Zero-based budgeting rarely succeeds in cutting costs.[32] Its real effect, in Musk's hands, was the concentration of power. His approach assumed that all expenditures were waste, and that bad data—whether fraudulent contracts, useless staff, or illegitimate people—could simply be deleted. What DOGE sought to automate, media researcher Eryk Salvaggio noted, was "not paperwork but democratic decision-making."[33] Efficiency became the alibi for centralization.

This centralization took material form in DOGE's approach to data, which aimed to put all of the government's information into a single repository. Washington had pursued the dream of data integration since the PATRIOT Act; in March 2025, Trump gave another push in this direction with an executive order about "Stopping Waste, Fraud, and Abuse by Eliminating Information Silos."[34] But the kind of total digital unification envisioned by DOGE was unprecedented. It found its most ambitious expression in the attempt to build a "mega API" at the IRS, which would make all taxpayer data—including names, addresses, Social Security numbers, tax returns, and employment information—accessible from one portal.[35]

The analogy to Silicon Valley platforms was deliberate. Uber had its "God view," letting employees watch every ride in real time. When Musk was buying Twitter, he demanded access to the platform's "firehose"—the unfiltered stream of all user activity.[36] Now the same principle was being applied to the state.[37] Palantir was a major partner in this undertaking. The company received over $113 million in government contracts in the first months of the Trump administration for work that included making it easier to integrate information from different agencies.[38]

Pooling data meant eliminating the legal and privacy guardrails that existed throughout the federal government. Silos are not necessarily bad things. They are spaces of privileged information. The barriers between them can be safeguards—checks against overreach, misuse, and surveillance.[39] But from the perspective of DOGE, they were obstacles to integration.

Such integration also served the purpose of facilitating the wider introduction of AI software, another priority for DOGE. To train AI models, and to use them to replace federal workers, data needed to be centralized and standardized. At the Department of Veteran Affairs, DOGE deployed an AI script to cancel unnecessary contracts. (The model hallucinated, mistaking contracts worth thousands for those worth millions of dollars.[40]) DOGE also used AI to locate diversity, equity, and inclusion (DEI) language in government policies and programs.[41] Most startlingly, in July 2025 DOGE's operatives announced the rollout of the "DOGE AI Deregulation Decision Tool," which they promised would cut 100,000 federal regulations within six months. In particular, they vowed to save 93 percent of the human labor involved in eliminating regulations by automating away the most time-consuming aspect: namely, reviewing comments submitted by American citizens. They boasted that hundreds of thousands of comments could be analyzed by AI almost instantly.[42]

DOGE's endpoint was governance by AI: the state not as a space of deliberation but as lines of executable code. As the sociologist Zeynep Tufekci observed, Musk cast himself as the "sysadmin" keeping the servers running.[43] He reinforced the conceit with a "Tech Support" T-shirt at cabinet meetings, presenting his role in apolitical terms. But the project was deeply political. DOGE's dream of data omniscience went beyond cost-benefit analysis or software modernization—those had been mantras of earlier administrations. For DOGE, the hunt for "waste, fraud, and abuse" blurred seamlessly into the hunt for illegitimate people: irregularities to be deleted. Muskism was not just about trimming budgets. Scaled to society, it meant purging those deemed out of place.

Shadow people

Once governance became a question of code, the next issue was obvious: which data counted as valid and which should be deleted? For Musk, the bugs were not only wasted dollars or redundant staff, but suspect people. Early in his DOGE tenure, he insisted that Social Security checks were going out to the dead—a conclusion born of misreading agency data.[44] Lacking experience in government, his team often struggled to interpret its systems. Musk joked on X, "Maybe Twilight is real and there are a lot of vampires collecting Social Security."[45] When asked by an interviewer to respond to critics like Bill Gates, who claimed that cuts to USAID would cost millions of lives, he dismissed them: "They don't even try to come up with a show orphan."[46] In his coder's idiom, "the empathy exploit" was simply "a bug in Western civilization" to be patched.[47] This had been integral to Musk's thinking for decades. His brother Kimbal took up the smartphone game *Polytopia* because Musk said "it would teach me how to be a CEO like he was." The first lesson was "Empathy is not an asset." The second was "Play life like a game."[48]

Treating life like a game had its own ethos and its own philosophers. In a theory often cited by Musk, Nick Bostrom speculates that we may be living in a simulation running on a mainframe in the future.[49] Further, many of the people around us may not be human beings but computer programs: what Bostrom calls "shadow-people," convincing imitations that lack interiority.[50] The ethical consequences are significant. If we are surrounded by shadow people, then appeals to empathy are not moral imperatives but manipulative code. The rational response is to steel yourself against humanitarian sentiment. The economist Robin Hanson came to this conclusion in a famous article called "How to Live in a Simulation" in 2001. "If you might be living in a simulation," he wrote, "then all else equal it seems that you should care less about others."[51]

Shadow people have been a bright line through Musk's career. At PayPal, they were fraudsters with false identities; at Twitter, first bots, then the far-fetched notion of "ghost employees"—the belief that many of the people on the payroll were not actually real humans.[52] Musk also expressed this idea through the concept of the NPC. Biden was "an NPC with a limited dialogue tree";[53] the press a "hivemind"[54] or "drone collective"[55] of "NPC media puppets."[56] He shared the meme of a head being opened to swap out a chip reading "Tesla good" for one reading "Tesla bad." The caption: "New program for the NPCs."[57] "Individuals should always wonder who wrote the software running in their head," he commented.[58] "Most humans have very limited firewalls," he elaborated elsewhere, "so are easily programmed."[59]

Seeing the world as code bled easily into politics. Musk called George Soros a "system hacker" who was funding a "fake asylum-seeker nightmare" while NGOs were bankrolling "fake protests" against Tesla dealerships.⁶⁰ The federal government was overrun with fraud, Musk claimed. It belonged to a broader exploit: Democrats were using the money to import undocumented immigrants en masse to "create a permanent majority—a one-party state."⁶¹ He believed they were doing so through a "hack" of asylum law.⁶² "Just say the magic phrase 'I seek asylum' and you're in," he said, "no evidence at all is required."⁶³

According to Musk, Biden had opened the border.⁶⁴ The United States was "rolling out the red (in more ways than one) carpet for homicidal cannibals" and declared that "illegals" could vote—meaning that "2024 will probably be the last election actually decided by U.S. citizens."⁶⁵ The day before the election, he told the podcaster Joe Rogan and his tens of millions of subscribers that migrants were "literally being flown into swing states," in some cases resulting in "700% increases" in the number of undocumented residents.⁶⁶ The border, he said, "basically doesn't exist."⁶⁷

These statements were not true. The border was not open. Asylum seekers were vetted, and many applications were denied. No non-citizens, let alone undocumented people, were permitted to vote and incidents of such fraud were vanishingly rare. There were no cannibals. An analysis of tens of thousands of tweets by Bloomberg found that Musk had become "X's biggest promoter of anti-immigrant conspiracies." In 2024, he tweeted over 1,300 times about immigration and voter fraud, receiving about 10 billion views in total.⁶⁸ An analysis run by the *Economist* found that, by 2024, Musk was posting about immigration almost twice as often as his second-favorite topic, free speech.⁶⁹

Musk's panic-mongering about people out of place expanded beyond the United States. These were the same months he promoted the European far-right demand for the forced "remigration" of its immigrant population. Remigration was the human equivalent of zero-based budgeting: wipe the slate clean, remove redundant or illegitimate entries, and start over. The vehemence of his anti-immigrant sentiment must be seen alongside his cybernetic view of society and the state. Marko Elez, a member of DOGE who was given write access to federal payment systems, put this pithily when he posted that 99 percent of Indians on temporary work visas "will be replaced by slightly smarter LLMs, they're going back don't worry guys."[70]

The convergence of code and nativism was stark. DOGE's most consequential act of data integration was designed to counteract the (imagined) incursion of "imported" migrants by accelerating mass deportations. By March 2025, Musk's operatives had begun building what *Wired* called a "master database" to track immigrants—knitting together records across the Department of Homeland Security (DHS), IRS, Social Security Administration (SSA), and voting rolls.[71] It dovetailed with Palantir's $30 million "ImmigrationOS" contract with Immigration and Customs Enforcement (ICE), which promised "near real-time visibility" on non-citizens.[72]

The next month, the *New York Times* reported that the Trump administration was using DOGE's new data integrations to render people illegible in the system. Thousands were added to the SSA's "death master file," which cut off access to their credit cards and bank accounts. One former commissioner called it "financial murder."[73] The goal was to choke out people's ability to make a living and force them to "self-deport." Trump also called for a national census that would not count undocumented residents, breaking with the principle that had guided American practice since 1790—the exceptions being enslaved people, counted as 3/5 of a person, and indigenous people, not counted at all.[74] What began as "tech support" for government databases meshed seamlessly with exclusionist politics.

This was a digital-native nativism. The debugging metaphor had become literal.

The red pill

Before and during his time at DOGE, Musk repeatedly referred to what he was doing as "reprogramming the Matrix."[75] But what did he mean? In the film, a young hacker named Neo (played by Keanu Reeves) discovers that life is a simulation: humans are stored in bio-mechanical honeycombs as batteries for a digital superintelligence that pacifies them with an illusion. "Take the red pill," Musk tweeted for the first time in May 2020, invoking the scene where Neo must choose between seeing the truth (red pill) or staying in blissful ignorance (blue pill).[76] The phrase has become a byword for the twenty-first-century far right, popularized on the subreddit r/TheRedPill, founded in 2012.[77] Neo's transformation from passive observer to active combatant is catalyzed by the red pill, which makes him traceable by his allies. Taking the red pill is not just about seeing through the simulation but mastering it. The decisive shift comes when Neo learns to reprogram the Matrix from the inside, enabling him to defeat the black-suited agents who hunt down anomalies.

By the end of *The Matrix*, Neo has adopted a chilling interpersonal style. If everyone is living in a simulation, then nobody's life matters. The film's most dramatic sequence features Neo in an overcoat doing cartwheels while mowing down dozens of enemies with guns in both hands and a blank expression on his face. But these are not acts of murder any more than they would be in a video game. The casualties are fake people.

The Matrix is a touchstone for the manosphere, an online community proudly touting male supremacy and misogyny. Musk has sometimes been described as being part of this group.[78] He has indeed been welcomed by manosphere leaders like Andrew Tate, whose Twitter account Musk restored. In an online conversation with Musk, Tate said the purchase of the social media platform had "cracked the Matrix in real time and it becomes extremely difficult now to run the psyops they were previously running and enslave the populace which is their primary goal."[79] Yet there is a basic but important difference between Musk and the manosphere. Tate talks about the need to *break* the matrix or *escape* from it. Musk is the rare figure who sees the matrix not as the problem, but as part of the solution. Muskism, in the end, means building a better matrix.

This is what Musk was trying to do with DOGE, combining ludic qualities of gaming with the fear of infiltration and an attempted renovation of governance through coding. In April 2025, he made the analogy literal, posting an image of himself as Neo brandishing two submachine guns in the lobby of the CIA.[80] But reprogramming the Matrix was not as easy as he imagined. Musk exited DOGE at the end of May 2025 after his 130-day tenure as a "Special Government Employee" not in sunglasses but with a black eye (reportedly delivered by his toddler). By then, DOGE's website claimed $170 billion in savings. An investigation by the *Financial Times* could verify "only a sliver of that figure."[81] After the generous tax cuts for the wealthy in Trump's One Big Beautiful Bill were passed in July 2025, the savings were erased altogether.[82] DOGE became what Musk described as a "whipping boy" for all manners of public discontent.[83] Tesla sales plummeted along with Musk's popularity.[84]

★

Seen one way, DOGE exposed the limits of Muskism as a mode of governance. Companies can treat workers as disposable units because the surrounding state guarantees their basic existence. Musk had deleted workers in his own companies ruthlessly and made deft use of labor law's loopholes, but in seeking to make real cuts at DOGE, he collided with the fragile biopolitical contract at the core of American life—misleadingly called "entitlements," but better understood as the survival infrastructure for many millions of people. As Americans vented their anger at feared or actual loss of access to Social Security and Medicare benefits, Musk's reputation suffered. The bond between Musk and his reply-guys failed to scale into a social contract.

Musk had imagined DOGE as the realization of the dream of reactionary technocracy, in which engineers disciplined society like a factory floor. But society is not a factory. It encompasses children, the elderly, the disabled, the geographically stranded—the very categories of life that markets define as surplus. In trying to impose a cyborg logic of optimization, Musk discovered that humans were not programmable units, and that the public sector's role is precisely to provide goods that the private sector can't or won't. The conflation of codebase, company, and state didn't work.

From another angle, of course, it was the state that instrumentalized Musk. Trump's circle used Musk to wage war on the "woke" domains of higher education, foreign aid, and scientific research under the sign of efficiency while attacking the administrative state and terrorizing the federal workforce in ways broadly consistent with the goals of conservative and libertarian thinktanks. DOGE also gave an experimental prod to the flesh of the welfare state to see how the body politic responded. When resistance surged, Musk absorbed the blame, but many of the changes remained. Perhaps the most important was the federal government's expanded capacity for domestic surveillance, as facilitated through DOGE's data integration efforts. In this sense, Muskism proved not a governing philosophy but a toolkit available to those who govern.

The denouement can thus be read in two registers. One is personal: Musk's overreach, his inability to transform celebrity capital into co-leadership of the state. The other is structural: the deeper implantation of contractors like Palantir into the back end of government. For all the pyrotechnics of the feud with Trump that unfolded in the months following Musk's departure from Washington, Musk's companies continued to be awarded new Pentagon contracts. In July 2025, xAI announced that it had entered a contract with the Department of Defense for its "Grok for Government" suite of AI products.[85] This happened the same day that the company introduced "Ani," an animated AI companion that can engage in sexually explicit conversations. She is an anime character with exaggerated bodily proportions and a childlike face—a textbook "waifu" reminiscent of the character Joi from the 2017 film *Blade Runner 2049*. This resulted in the improbable but symptomatic headline: "Grok rolls out pornographic anime companion, lands Department of Defense contract."[86]

In the 1990s, futurists saw the coming of "cyberstates." New technologies would allow companies and communities to dissolve the government and govern themselves. Three decades on, it seems that political and digital power can be symbiotic. Under Muskism, tech came not to bury the nation state but to enhance it.

Conclusion:
Four Futures for Muskism

In *The Great Transformation*, published in 1944, the Hungarian economist Karl Polanyi described the world market as coming to resemble "a gargantuan automaton."[1] The peoples of the Earth had been yanked out by the root from their places of origin and turned into movable units, dragged across oceans and moved from the countryside into cities as labor power. The resources and materials of the world were being extracted from the soil and recombined across great distances—refined, melted, nailed, and bolted together into new forms. Payment systems pushed money through the automaton's piping, stabilized by a gold standard managed by dials and levers in the financial citadels of London and New York. Globalization had assembled people, products, and places into a kind of patchwork humanoid. Elsewhere, Polanyi used a metaphor from his own Jewish tradition: the market was a golem created by humans but now slipping from their control, threatening to destroy the Earth itself. The human race had become "a tortured soul" looking at a "terrible machine."[2]

In the third millennium, it is still easy to see global capitalism this way—as a kind of cyborg we have created collectively but which now threatens our survival through climate breakdown and the still-looming threat of thermonuclear war. But the golem has evolved. To the industrial products of Polanyi's world—the automobiles, steel girders, steamships,

CONCLUSION

and trains—we have added computers, rocket ships, fiber optic cables, satellites, drones, and smartphones. By the twenty-first century, we have a new name for the golem—the mech. What face can we put on the mech of our times? There is one clear candidate. No person embodies the twenty-first-century man-machine more than the subject of this book, Elon Musk.

In 2021, in a move still reflected on the company's website, Musk changed his official title at Tesla from CEO to something new: Technoking. In 1651, Thomas Hobbes published *Leviathan*, with a frontispiece depicting a sovereign holding a scepter in one hand and a sword in the other, his body composed of his subjects. One can imagine Musk arrayed in a similar fashion: a torso made of a Tesla chassis, a SpaceX rocket in one hand, a Neuralink at the temple, logged on to X. In fact, you need not merely imagine this—you can generate a realistic image of it in under fifteen seconds through Grok.

Where will the Muskiathan move next? We can imagine four futures.

The first could be called *Carbon Musk*. It is not beyond the realm of possibility that Musk finds his way back to the field of engagement that made him most famous, namely, as a pioneer of electrification. Whether or not one agrees that electrifying the personal consumer vehicle is the best path to a decarbonized future, the Tesla ecosystem could materially improve people's lives. Having solar-generated battery storage for backup in times of deadly or merely inconvenient weather would be a net benefit.

The XPRIZE Foundation has been one of the largest of the very few recipients of Musk's philanthropy. He pledged $112 million for the best idea for carbon removal.[3] One can imagine Musk making a hard charge into geoengineering and similar technologies to transform the earth's climate to make it more hospitable to human life. If critics have long pointed out that even a radically degraded planet is still far more comfortable than any version of a terraformed Mars, maybe Musk will come to this sensible conclusion too.

At the same time, Musk has grown ever more determined to position Tesla as a robot company. The central image on the cover of the company's fourth "Master Plan" is an Optimus robot stowing a bag of groceries.[4] Musk predicted that 80 percent of the company's profits would come from rolling out Optimus at scale—making his usual outlandish claims despite having failed to sell a single commercial unit.[5]

Musk has also prioritized artificial intelligence, imagining millions of Teslas as a "giant distributed inference fleet," using idle computing capacity to run AI models.[6] He predicts that, in time, the energy of the sun—and later, that of an entire galaxy—will be harnessed to power intelligent machines.[7] The ecological costs of AI are well known—and distressingly evident in the air pollution produced by xAI's facility in Memphis. The energy usage of data centers will soon rival those of cities and even countries. The researcher Kate Crawford calls AI a metabolic technology that is "eating the future."[8] Carbon Musk and the mass production of robots are mutually exclusive. Fulfilling his most extreme cyborg dreams will be at the expense of a livable planet.

There is also a version of Musk's future that delves deeper into state symbiosis. Call it *Contractor Musk*. If becoming a public official with DOGE had limited returns, then perhaps it would be easier to withdraw to the more traditional role of private actor, provisioning the needs of the state. The co-founder of Palantir, Alexander Karp, laid out a blueprint for such a role in his book *The Technological Republic*. Karp touches on some of the subjects covered in this one: the origin of Silicon Valley in government funding, the ways that networks came to influence both military and tech-world thinking, and the new business opportunities being created by a battlefield increasingly defined by cheap, software-driven drones. Karp is critical of Silicon Valley's focus on convenience-oriented consumer apps and wants it to return to its roots as a Pentagon contractor, albeit in a more sophisticated, nimble, and entrepreneurial form than military-industrial incumbents like Northrop Grumman and Lockheed Martin.[9] Palantir has seen its valuation soar during the second Trump administration, as it sells data integration and analytics services to agencies throughout the state.

Trump's announcement of the so-called Golden Dome missile defense project opens new vistas for military contractors. It closely echoes the Star Wars program of the Reagan era while pointing to a further militarization of space. SpaceX landed a $2 billion contract with the Pentagon to develop satellites for Golden Dome, while Musk's "Project Starfall" is exploring the idea of using Starships to transport military material around the globe.[10]

Meanwhile, Musk's embrace of right-wing politics has led him ever further into the language of violence. In June 2025, he posted a slightly cryptic tweet that "Whatever happens, we have got the spaceships, and they do not."[11] The quote is from the author and politician Hilaire Belloc, who in 1898 wrote the couplet: "Whatever happens we have got/The Maxim gun, and they have not."[12] This was the first fully automatic machine gun, which was used to great effect in colonial wars. In the words of one historian, it "turned battles into one-sided massacres."[13] In September 2025, Musk spoke by video link to over 100,000 people gathered at the Unite the Kingdom rally organized by the far-right activist Tommy Robinson in London. Denouncing the "rapidly increasing erosion of Britain with massive uncontrolled migration," Musk told the crowd, "Whether you choose violence or not, violence is coming to you. You either fight back or you die."[14]

Another scenario consistent with Musk's recent direction of travel is one we could call *Compound Musk*. This dovetails with a leitmotif of Muskism that has received scant attention in this book so far: demographic decline. "If people don't have more children, civilization is going to crumble," Musk warned in 2021 in one of countless similar remarks.[15] By 2024, he was declaring "extreme birth rate collapse is the biggest danger to human civilization by far."[16] At a right-wing youth festival in Rome in 2023, he claimed the global population would shrink to one tenth its size within three generations, when in fact it is projected to rise to 10 billion. He said birth rates were "maybe half the replacement rate" when globally they remain just above replacement.[17]

At first glance, Musk's concern seems pitched at the scale of the species. Yet he consistently highlights fertility in certain regions while ignoring others. In May 2025 he shared a world map of birth rates labeled "The Great Population Collapse," eliding the fact that sub-Saharan Africa still shows high fertility.[18] Put simply, Musk's demographic panic is bound up with his concern for the survival of white civilization. In his enthusiastic support for far-right parties in Europe and the UK, he has stated this even more plainly. "White people are a rapidly diminishing minority of global population," he posted in September 2025.[19] "Low birth rate is the number one threat to the West followed closely by migration," he posted the same month. "There will be no West if this continues."[20] The problem was not how many people were having kids but *which* people and *where*.[21]

This rhetoric has ample precedent in Musk's native South Africa, where white politicians in the apartheid era warned of being "swamped" and sought to win the "cradle race" by boosting white fertility.[22] In 2025, Musk amplified Afrikaner extremists on social media alleging to be victims of a "white genocide" perpetrated by South Africa's Black majority. Their claims, promoted through his platform, reached the Oval Office, where Trump confronted South African president Cyril Ramaphosa with false evidence of anti-white crimes. Changes to U.S. refugee policy were already underway: Afrikaners were admitted while other asylum programs were frozen.[23] The invocation of the "genocide" of whites has been used in South Africa since at least the 1960s—as a way to criticize white women aborting fetuses. For white South Africans, the prospect of a shrinking white population was interpreted as evidence of "race suicide."[24]

Musk has tackled the problem of demographic decline personally, having fathered at least fourteen children. Of these, the first seven were assigned male at birth—a statistical improbability that could be explained by proactive sex selection before embryo implantation. Musk adopted the same approach to educating his children as he did to rockets and cars: bring the process in-house. Dissatisfied with Los Angeles private schools, in 2014 he founded his own school—called "Ad Astra"—inside the SpaceX factory.[25] As his family grew, so did his bespoke institutions. Schools followed his relocations to and within Texas: first Brownsville in 2021, then Bastrop in 2024, each positioned beside his factories and embryonic company towns.[26] Where Henry Ford had built new towns in the Amazon rainforest, Musk was building them inside the border walls of the nation.[27]

The compound approach to education aligned with the parallel venture of Musk's Silicon Valley peers. These included "Bezos Academy" preschools in Texas and Palantir co-founder Joe Lonsdale's University of Austin. In 2023, Musk himself seeded a new primary and secondary science school with $100 million.[28] Texas governor Greg Abbott hailed it with a tweet:

"Musk University > Harvard University."[29] Musk floated a more juvenile name: "Texas Institute of Technology & Science"—TITS.[30]

Meanwhile, Musk began spending time with self-proclaimed "pro-natalists" Simone and Malcolm Collins. Malcolm's brother, who ran a fertility clinic, ended up joining DOGE.[31] The Collinses speculated in 2022 that a billionaire who perfected artificial wombs could repopulate civilization with his own DNA.[32] Their own children bore names like Titan Invictus and Industry Americus. The resonance with Musk was obvious. When right-wing influencer Ashley St. Clair announced that she and Musk had a child named Romulus—the mythical founder of Rome—the fantasy of bloodline empire-building was made explicit. Roman models of paternal lineage haunt Muskism: decline will be reversed by a new imperial line. At one point, Musk texted St. Clair, "To reach legion-level before the apocalypse, we will need to use surrogates."[33]

Compound Musk shrinks the horizons of humankind into the defense of white fertility, and further still to the confines of the private settlement, where children are raised outside public schools. A future once mapped onto the galaxy resolves into a ring-fence around the home. Vast and claustrophobic at once, Compound Musk is where the cosmic scale of Muskism collapses into the fruit of many wombs nurtured by the breadwinning fortunes of one patriarch.

A fourth future, picking up on the last half of this book, might be called *Cyborg Musk*. When he made his chainsaw-wielding appearance at the Conservative Political Action Conference in early 2025, Musk wore a T-shirt that said, "I'm not procrastinating, I'm doing side quests."[34] At that point, DOGE was a month old. Several months later, after leaving the government, Musk explained at a public event that he had found

Washington, perhaps predictably, too political. It was an "interesting side quest," he said, but now he was returning to the main quest, which he described as "pedal to the metal on humanoid robots and digital superintelligence." He outlined a future of what he called our "future machine descendants, or mostly machine descendants," laying his cyborg cards on the table.[35]

Musk predicted that within a decade there may be as many as ten times the number of humanoid robots on Earth as humans. He foresaw the arrival of digital superintelligence by 2026 at the latest. By this measure, he reflected later, the efforts of DOGE were like cleaning a beach of "needles and feces and trash . . . But then there's also this thousand-foot wall of water which is a tsunami of AI. And how much does cleaning the beach really matter if you've got a thousand-foot tsunami about to hit?"[36]

By this reading, Musk's time in government, interpreted by many as his attempt to defeat the final boss of the state, appears in a different light. If he believes that digital superintelligence will arrive within a year or two, and that humanoid robots will outnumber humans ten to one within a decade, then two things follow: first, he needs to be in charge. In late 2025, he urged Tesla's shareholders to approve a measure that could hugely increase his ownership stake. "My fundamental concern with regard to how much voting control I have in Tesla," he said, "is if I go ahead and build this enormous robot army, can I just be ousted at some point in the future?"[37]

Second, human rule itself is impermanent. DOGE is less a lasting fix for the state's inefficiencies than an interregnum project: a way of stabilizing the human-administered bureaucracy long enough for the "tsunami" of AI to arrive. The point is not to perfect government for the long run, but to clear its ledgers, rationalize its code, and make it legible to the algorithms and automatons poised for entry.

CONCLUSION

What would the ideal bureaucrat of future Muskism look like? Imagine a day in 2035. He arrives to work in a Starlink-enabled fully self-driving Cybertruck with a Neuralink implant. Seated at a terminal with multiple suspended monitors, he ports with Grok—now integrated via a "mega API" across all government databases. An AI agent alerts him to an undocumented immigrant spotted on a street corner in Boston by a camera equipped with facial recognition technology. The bureaucrat routes the notification to ICE, which dispatches a team of heavily armed officers along with a humanoid robot backup. The arrest is recorded by a drone, the footage subsequently synced to a lo-fi electronic track and posted on government YouTube channels. Data is also sent to a games developer working on a first-person shooter about chasing down illegal aliens. The bureaucrat sips coffee served by a personal Optimus, its snap-on face customized to his liking.

The job of the Muskist bureaucrat is to tune the signal-to-noise ratio of a giant cybernetic organism called the United States. Grievances are handled via hotline by animated AI bots adapted to the demographic profile of the constituent. "Sexy mode" is an option. The bureaucrat is already half ready to be replaced, and this is the point. The real endgame is to make himself redundant, to leave a system so rationalized that future cyborg humans can inherit it without friction. Under Muskism, the civil servant is a transitional species. State reform is the bridge to a new regime of algorithmic sovereignty, in which consent and legitimacy are displaced by the consistency of machine rule.

Now imagine a day in the life of the bureaucrat's daughter. The scenario that follows is built entirely on Musk's own statements. She arrives at school, passing by the mural of a young blond Ukrainian woman murdered by a mentally ill Black man.[38] In a lesson titled "The Rape of Europe,"[39] she is told about a

CONCLUSION

dark time in the early twenty-first century when countries like the United Kingdom were "invaded" by people from non-Western countries welcomed in by deluded politicians practicing "suicidal empathy." This was remedied by the round-up and deportation drives of the 2030s carried out under the slogan of "generation remigration."[40]

The "defense of Western Civilization," she is told, was a success and parties once called "far right" are now firmly in power. Their sovereignty is secured by the Starlink communication grid of satellites and networks of rockets capable of carrying hypersonic missiles. Jobs exist for some high-level managers and programmers at automated mines and factories, but most people spend their recreational hours streaming video, scrolling and posting on social media, and gaming. This is an immersive and endlessly diverting experience because everyone is now equipped with a Neuralink that expands the bandwidth of one's interface with the "collective AI" to a rushing river of data while enhancing the senses, letting each person see in ultraviolet like a dragonfly and infrared like a vampire bat.

After the daily moment of quiet reflection on the importance of curiosity and the vastness of the universe, students port their Neuralinks into Grokipedia modules on history. They learn that the term "slave" originally referred to white people and the British Empire was important mostly for ending African slavery.[41] Terms like "racism" and "sexism" never appear as they have been proven to correlate directly with the "woke mind virus"—considered eradicated since 2032.[42] The great error of history in the dark years of the early twenty-first century was that "weak makes right."[43] This was part of what is described as the "vampire cult" of education now mercifully superseded.[44] The powerful, she learns, should never apologize.

Even great civilizations, though, do not last forever. She

absorbs the surprising fact that all empires have fallen for the same reason: low birth rates.[45] Fewer people mean not only less labor power—no longer relevant due to the replacement of almost all workers by humanoid robots—but, more importantly, less human intelligence. As she has been taught many times, intelligence is the "the primary evolutionary vector for humans."[46]

On guard against further decline, her science class teaches her about the latest developments in selective human breeding for higher intelligence, including "iterated embryo selection" which allows for choosing specific genetic alleles and cross-fertilizing them without implanting them in a womb, as well as the now-standard "spell-checking" of genetic code to remove all mutations.[47] Learning about this technology is framed as a patriotic duty to ensure strong offspring and reduce future burdens on the privatized health system. It is incumbent on smart women like her to produce at least 2.1 children.[48] She is reminded again of the self-evident truth that there are only two genders and of the evils of past decades, when people changed their genders in what is now recognized as "child mutilation and sterilization." Such actions are now punishable by imprisonment.[49]

More important of course is the pursuit of digital super-intelligence. Advances have already made possible the creation of a fleet of humanoid robots ten times the human population on earth. They are a constant presence on streets and in homes. The school itself is maintained and guarded by them. They also lead the gym class, which doubles as border defense education with drills and carbon-fiber weapons.

On the way home, the air is hazy with vapor and filled with a low-level hum from the cooling towers of the Neural Districts—massive data center complexes ringing the city. A Tesla sign on a fence reads "Data is Sovereignty." Asking her parents why the trees are dead and the streambeds dry, they respond "critical infrastructure." At home, the lights are dim because "non-essential use" is rationed to supply the data centers at peak hours.

She talks to her parents about Mars. Starship is now launching twenty times a day and the rockets have become fully reusable, but the launch window only comes every two years. Promotional videos and gleeful simulations populated by animated companion AIs circulate constantly on X. One more push, one more sacrifice, they all sing, and humanity will ascend.

Acknowledgements

We're grateful to everyone who made this book possible. Our agents Mel Flashman and Molly Atlas were indispensable as always. Molly's colleagues Claire Nozieres and Zoe Willis helped our book find a home with foreign publishers. Our editors, Sarah Haugen at Harper and Tom Penn at Allen Lane, believed in this project from the beginning and pushed us to fulfill its potential. Richard Duguid at Penguin Press UK shepherded the manuscript through a complicated production process. Thanks to the marketing and publicity teams at both Harper and Allen Lane—Heather Drucker, Jessica Gilo, Matthew Hutchinson, Louis Cluzan, Gavin Read, and Olivia Kumar—for all their hard work to put our book into people's hands, as well as the efforts of Heinrich Geiselberger and Leonie Hohmann at Suhrkamp, Séverine Nikel and Séverine Roscot at Le Seuil, and many others we will meet soon. We deeply appreciate our translators for making it possible for us to reach readers throughout the world; if we're ever in the same town, we'd like to buy you a drink.

We also feel lucky to have such interesting and loyal friends, many of whom helped us think through the themes of this book. A partial list would have to include Kat, Jamie, Kirsten, Aaron, Matt G., Dan D., and Thea. Finally, we owe Chris Lydon and Mary McGrath a special debt of gratitude. They hosted and produced the episode of the *Open Source* podcast where the idea for this book was born.

Each of us has a few further acknowledgements to make individually.

Quinn would like to thank his editors at *New Statesman*, the *Guardian*, and the *New York Review of Books* for soliciting and

ACKNOWLEDGEMENTS

fine-tuning early versions of the arguments made here, as well as Amy Goodman and *Democracy Now!*, CBC *Ideas* and *Front Burner*, Joshua Citarella and *Doomscroll*, Eleanor Penny and Novara Media, Doug Henwood, John Ganz and Max Read for putting them in front of smart and politically engaged audiences. Quinn's friends have suffered through months of Musk minutiae for which they deserve extra gratitude and requests for forgiveness. Per decades of habit, Ryan and Hadji have been first sounding boards. Thanks also go to James D., Moira, Hank, Tom M., Pavlos, Stefanos, Simon D., Tina, Ana, Justin, Dylan, Will, and the chats: Research Group and Alex Trebek a.k.a. Lifestyles of the Rich and Miserable. He would also like to thank the John Simon Guggenheim Memorial Foundation for the time to write and to his colleagues and students at Boston University for the welcoming academic home and base from which to work and think freely. This book took all the struts of support and sources of joy available. Michelle and Yann were the most important ones, inspiring and assuring in turns, with Quinn's more geographically distant family warming him from afar. Quinn and Mayana are still waiting for Godot. He takes none of his extravagant good fortune for granted and thanks his lucky stars every day.

Ben would like to thank his editors at the *New York Review of Books* for commissioning and editing the pieces that gave him his first opportunities to grapple with Musk. He is especially grateful to Max Nelson. Ben's wife, Moira, is the foundation stone with whom everything is possible and without whom nothing is possible; he is grateful to her for all things, always, and to their three daughters, Zoe, Josephine, and Sylvia, who did not help him write this book. Thanks also to Moira's parents, Bill and Kathy, and to Ben's mother, Mathea. Ben's father, Peter, is no longer with us but he is reading these pages somewhere, chuckling and shaking his head.

Notes

INTRODUCTION

1. See Stefan Link, *Forging Global Fordism: Nazi Germany, Soviet Russia, and the Contest over the Industrial Order* (Princeton, NJ: Princeton University Press, 2020), 3–8.
2. Jill Lepore, "Elon Musk Is Building a Sci-Fi World, and the Rest of Us Are Trapped in It," *New York Times* (November 4, 2021), https://www.nytimes.com/2021/11/04/opinion/elon-musk-capitalism.html; Tom McTague, "Is Elon the new Enoch?," *UnHerd* (January 11, 2025), https://unherd.com/2025/01/is-elon-the-new-enoch/; Jill Lepore, *X Man: The Elon Musk Origin Story*, podcast audio2025. Harold Meyerson, "Ford and Musk. They Made Cars. They Backed Fascists.," *The American Prospect* (January 6, 2025), https://prospect.org/2025/01/06/2025-01-06-ford-musk-made-cars-backed-fascists/.
3. "Joe Rogan Experience #2281 – Elon Musk," (February 28, 2025), https://www.youtube.com/watch?v=sSOxPJD-VNo.
4. https://x.com/elonmusk/status/1953759666380714489, August 8, 2025.
5. https://x.com/elonmusk/status/1961349227487084829, August 29, 2025.
6. "Elon Musk interview with Seth Dillon, Kyle Mann, and Ethan Nicolle," (December 22, 2021), https://www.youtube.com/watch?v=jvGnw1sHh9M.
7. Viktor Valgarðsson et al., "A Crisis of Political Trust? Global Trends in Institutional Trust from 1958 to 2019," *British Journal of Political Science* 55 (2025), https://doi.org/10.1017/S0007123424000498.

PART ONE: FOUNDATION

1. Walter Isaacson, *Elon Musk* (New York: Simon & Schuster, 2023), 33; Ashlee Vance, *Elon Musk: Tesla, SpaceX, and the Quest for a Fantastic Future* (New York: Ecco, 2015). In February 2018, when Musk launched a Tesla Roadster into space aboard the SpaceX Falcon Heavy rocket, the car's glovebox contained an optical data disk with a copy of Asimov's *Foundation* trilogy.
2. Isaac Asimov, *I, Asimov: A Memoir* (New York: Random House, 2009), 117.

1. FORTRESS FUTURISM

1. Joshua Haldeman, "America Needs No Part of the Price System," *Technocracy Digest*, no. 37 (July 1940): 6.
2. Howard Scott, *History and Purpose of Technocracy* (1965), 22. https://archive.org/details/HistoryAndPurposeOfTechnocracy.howardScott.
3. Jill Lepore, "The Failed Ideas That Drive Elon Musk," *New York Times* (April 4, 2025), https://www.nytimes.com/2025/04/04/opinion/elon-musk-doge-technocracy.html.
4. E. Merrill Root, "The Culture of Abundance," *Technocracy Digest*, no. 79 (Jan 1945): 40.
5. Haldeman, "America Needs No Part of the Price System," 6.
6. "Seeking home by plane and car," *Liverpool Daily Post* (September 11, 1950).
7. Quoted in Joshua Benton, "Elon Musk's Anti-Semitic, Apartheid-Loving Grandfather," *The Atlantic* (Sep 20, 2023). For original text see "Loof Optrede van N.P.-Bewind: Kanadese Politikus Vestig Hom in S.A.," *Die Transvaler* (November 22, 1950), https://gpa.eastview.com/dtsa/newspapers/dtsa19501122-01.1.5.
8. In the same 1960 text, Haldeman quoted "the prophetic and emphatic statement" of an unnamed Anglican minister in Toronto: "South Africa will become the leader of white civilization in the world." Jill Lepore, "The World According to Elon Musk's Grandfather," *The New Yorker* (Sep 19, 2023), https://www.newyorker.com/news/daily-comment/the-world-according-to-

elon-musks-grandfather. See also Ira Basen, "In science we trust," *CBC.ca* (June 28. 2021), https://newsinteractives.cbc.ca/longform/technocracy-incorporated-elon-musk/; Geoff Leo, "The Canadian roots of Elon Musk's conspiracist grandpa," *CBC News* (March 20, 2025), https://www.cbc.ca/newsinteractives/features/joshua-haldeman-elon-musk-saskatchewan-tech-utopian-conspiracist.

9. Quoted in Hilton Judin, *Architecture, State Modernism and Cultural Nationalism in the Apartheid Capital* (New York: Routledge, 2021), 33.
10. Paul N. Edwards and Gabrielle Hecht, "History and the Technopolitics of Identity: The Case of Apartheid South Africa," *Journal of Southern African Studies* 36, no. 3 (Sep 2010): 625.
11. Jeevan Vasagar, *Lion City: Singapore and the Invention of Modern Asia* (New York: Pegasus, 2022), 225.
12. See Judin, *Architecture, State Modernism and Cultural Nationalism in the Apartheid Capital*.
13. Edwards and Hecht, "History and the Technopolitics of Identity: The Case of Apartheid South Africa," 621.
14. Glenn Adler, "From the 'Liverpool of the Cape' to 'The Detroit of South Africa': The Automobile Industry and Industrial Development in the Port Elizabeth-Uitenhage Region," *Kronos* 20 (Nov 1993): 37.
15. Stephen Gelb, "Making Sense of the Crisis," *Transformation* 5 (1987): 33–50.
16. Deborah Posel, "A mania for measurement: statistics and statecraft in the transition to apartheid," in *Science and society in Southern Africa*, ed. Saul Dubow (Manchester: Manchester University Press, 2000), 131–32.
17. Laura Evans, "Contextualising Apartheid at the End of Empire: Repression, 'Development' and the Bantustans," *The Journal of Imperial and Commonwealth History* 47, no. 2 (2019): 373.
18. Keith Breckenridge, *Biometric State: The Global Politics of Identification and Surveillance in South Africa, 1850 to the Present* (New York: Cambridge University Press, 2014).
19. Quoted in Edwards and Hecht, "History and the Technopolitics of Identity: The Case of Apartheid South Africa," 633.
20. R. Kelly Garrett and Paul N. Edwards, "Revolutionary Secrets: Technology's Role in the South African Anti-Apartheid Movement," *Social Science Computer Review* 25 (2007): 13–26. For

further reflections on Musk, computers and South Africa see Efthimios Karayiannides, "The Mine Dumps of Silicon Valley," *Africa is a Country* (October 14, 2024), https://africasacountry.com/2024/10/the-mine-dumps-of-silicon-valley.
21. Ahmed Areff, "SA Rugby Culture Creates People like Pistorius," *News24* (July 22, 2015), https://www.news24.com/sa-rugby-culture-creates-people-like-pistorius-elon-musks-father-20150722.
22. Vance, *Elon Musk: Tesla, SpaceX, and the Quest for a Fantastic Future*, 33.
23. Rachel Savage, "The making of Elon Musk: how did his childhood in apartheid South Africa shape him?," *The Guardian* (March 10, 2025), https://www.theguardian.com/technology/2025/mar/10/making-of-elon-musk-childhood-apartheid-south-africa.
24. Isaacson, *Elon Musk*, 12–13.
25. David Shandler, "Structural Crisis and Liberalism: A History of the Progressive Federal Party" (M.A. thesis University of Cape Town, 1991), 71–72. This was the Constitution of 1983, which was approved by white voters in a referendum on 2 November 1983.
26. Savage, "The making of Elon Musk: how did his childhood in apartheid South Africa shape him?"
27. Ibid.
28. John Eligon and Lynsey Chutel, "Elon Musk Left a South Africa That Was Rife With Misinformation and White Privilege," *New York Times* (May 5, 2022), https://www.nytimes.com/2022/05/05/world/africa/elon-musk-south-africa.html.
29. Isaacson, *Elon Musk*, 23.
30. Vance, *Elon Musk: Tesla, SpaceX, and the Quest for a Fantastic Future*, 44.
31. See e.g. Isaacson, *Elon Musk*, 31.
32. Douglas Adams, *The Hitchhiker's Guide to the Galaxy* (New York: Harmony Books, 1979), 30.
33. Vance, *Elon Musk: Tesla, SpaceX, and the Quest for a Fantastic Future*, 38.
34. Ibid.
35. Ibid., 33–4
36. djvlad, "Errol Musk, Father of Elon Musk, Tells His Life Story (Full Interview)," (February 13, 2025), https://www.youtube.com/watch?v=J5WyTw00XDs.

37. See Reina-Marie Loader, "Broadcasting Change: An Aerial Overview of South African Television Debates in an Age of Constant Transition," *Critical Studies in Television* 19, no. 1 (2024): 94–118.
38. "South Africa: The Other Vast Wasteland," *Time* (November 20, 1964), https://time.com/archive/6626982/south-africa-the-other-vast-wasteland/.
39. Quoted in Rob Nixon, "Apollo 11, Apartheid, and TV," *The Atlantic* (July 1999), https://www.theatlantic.com/magazine/archive/1999/07/apollo-11-apartheid-and-tv/377681/.
40. See e.g. TV listings, *Sunday Times* [Johannesburg] (September 7, 1980).
41. Matthew Bishop, "Do Panic: Elon Musk's Obsession with The Hitchhiker's Guide," *The Observer* (April 25, 2025), https://observer.co.uk/news/international/article/how-the-hitchhikers-guide-inspired-elon-musk; Parmy Olson, "Elon, Hold On to Your 'Star Trek' Dreams," Bloomberg (April 21, 2024), https://www.bloomberg.com/opinion/articles/2024-04-21/elon-musk-silicon-valley-should-hold-on-to-star-trek-dreams.
42. See e.g. listings in *Sunday Times* (May 31, 1988) for *Transformers* and (July 20, 1986) for *Robotech*.
43. *Sunday Times* (May 31, 1988).
44. *Sunday Times* magazine [Johannesburg] (February 22, 1987).
45. Tim Levin, "Elon Musk Wants to Give Amputees Robotic Limbs Powered by Chips Implanted in Their Brains," *Business Insider* (July 20, 2023), https://www.businessinsider.com/elon-musk-optimus-tesla-robot-limbs-neuralink-cyborg-2023-7.
46. https://x.com/elonmusk/status/1051389235406598144, October 14, 2018.
47. André Carl Van der Merwe, *Moffie* (New York: Europa Editions, 2011), 19.
48. https://www.saha.org.za/ecc25/bothas_emergency_legalized_murder.htm.
49. https://www.saha.org.za/ecc25/free_us_from_the_call_up.htm.
50. https://www.saha.org.za/ecc25/he_doesnt_look_like_a_terrorist.htm.

51. "Elon Musk Captured by Rainn Wilson! | Metaphysical Milkshake," (March 18, 2013), https://www.youtube.com/watch?v=NsoIHCj2q-E&t=31s.
52. Mark Gimein, "Fast Track," *Salon.com* (August 17, 1999), https://www.salon.com/1999/08/17/elon_musk.
53. Isaacson, *Elon Musk*, 38–41.
54. "CHM Revolutionaries: An Evening with Elon Musk" (January 22, 2013), https://www.youtube.com/watch?v=AHHwXUm3iIg.
55. Eve Fairbanks, "What Elon Musk Gets Wrong About South Africa," *The Dial* (November 30, 2023), https://www.thedial.world/articles/news/issue-10/elon-musk-walter-isaacson-south-africa-myths.
56. Vivian Chenxue Lu and Nana Osei-Opare, "Elon Musk Wanted the Cybertruck to Look Like 'the Future.' But It Reminds Us of One Particular Past," *Slate* (March 15, 2025), https://slate.com/news-and-politics/2025/03/tesla-cybertruck-protests-vandalism-elon-musk.html.
57. https://x.com/elonmusk/status/1903556327290626165, March 22, 2025; Khanyisile Ngcobo, "Claims of White Genocide 'Not Real', South African Court Rules," *BBC* (February 25, 2025), https://www.bbc.com/news/articles/cwyj1198wy3o.

2. THE SUPERSET

1. John Perry Barlow, "A Declaration of the Independence of Cyberspace," (8 Feb 1996), https://www.eff.org/cyberspace-independence.
2. Stephen C. Mooney, "The Cyberstate" (Proceedings of the 1996 ACM SIGCPR/SIGMIS conference on Computer personnel research, Denver, Colorado, USA, Association for Computing Machinery, 1996); James Dale Davidson and William Rees-Mogg, *The Sovereign Individual: How to Survive and Thrive During the Collapse of the Welfare State* (London: Macmillan, 1997).
3. Vance, *Elon Musk: Tesla, SpaceX, and the Quest for a Fantastic Future*, 76.
4. Gerd Häusler, "The Globalization of Finance," *Finance & Development* 39, no. 1 (March 2002), https://www.imf.org/external/pubs/ft/fandd/2002/03/hausler.htm.

5. Edward J. Malecki and Hu Wei, "A Wired World: The Evolving Geography of Submarine Cables and the Shift to Asia," *Annals of the Association of American Geographers* 99, no. 2 (Apr 2009): 370.
6. Bureau of Labor Statistics, "Computer ownership up sharply in the 1990s," *The Economics Daily* (April 5, 1999), https://www.bls.gov/opub/ted/1999/apr/wk1/art01.htm.
7. What was measured was whether they had been online in the previous three months. "Data Page: Share of the Population Using the Internet," part of the following publication: Hannah Ritchie, Edouard Mathieu, Max Roser, and Esteban Ortiz-Ospina (2023), "Internet." Data adapted from International Telecommunication Union (ITU), via World Bank. Retrieved from https://archive.ourworldindata.org/20250909-093708/grapher/share-of-individuals-using-the-internet.html.
8. https://www.internetlivestats.com/total-number-of-websites/
9. "Elon Musk: Elon Musk's Vision for the Future [Entire Talk]," (October 7, 2015), https://www.youtube.com/watch?v=SVk1hboZOrE.
10. Kevin Kelly, *Out of Control: The New Biology of Machines, Social Systems, and the Economic World* (New York: Basic Books, 1992), 215.
11. Fred Turner, *From Counterculture to Cyberculture: Stewart Brand, the Whole Earth Network, and the Rise of Digital Utopianism* (Chicago: University of Chicago Press, 2006), 149.
12. See Mariana Mazzucato, *The Entrepreneurial State: Debunking Public vs. Private Sector Myths* (New York: PublicAffairs, 2015).
13. Ben Tarnoff, *Internet for the People: The Fight for Our Digital Future* (New York: Verso, 2022), 17–18.
14. Dan Schiller, *Digital Capitalism: Networking the Global Market System* (Cambridge, MA: MIT Press, 1999), 144.
15. "'I don't even have a modem': An Interview with William Gibson," (November 23, 1994), https://josefsson.net/gibson/.
16. "Mosaic Launches an Internet Revolution," *U.S. National Science Foundation* (April 8, 2004), https://www.nsf.gov/news/mosaic-launches-internet-revolution.
17. David Einstein, "Netscape Mania Sends Stock Soaring," *San Francisco Chronicle* (August 10, 1995).

18. Vance, *Elon Musk: Tesla, SpaceX, and the Quest for a Fantastic Future*, 60–63; Isaacson, *Elon Musk*, 61–2.
19. Ibid., 66–71; Isaacson, *Elon Musk*, 63–5; "CHM Revolutionaries: An Evening with Elon Musk."
20. Clayton M. Christensen, *The Innovator's Dilemma: When New Technologies Cause Great Firms to Fail* (Boston, MA: Harvard Business School Press, 1997).
21. Isaacson, *Elon Musk*, 66.
22. Vance, *Elon Musk: Tesla, SpaceX, and the Quest for a Fantastic Future*, 64–5.
23. Bill Gates, *The Road Ahead* (New York: Viking, 1995).
24. Joelle Tessler, "Firms Look for Payoff from On-Line Payments," *Chicago Tribune* (July 17, 2000).
25. Isaacson, *Elon Musk*, 74.
26. "Top 10 Web Sites by Category," *Los Angeles Times* (May 1, 2000).
27. Isaacson, *Elon Musk*, 83–5.
28. Gary Wolfe, "The (Second Phase of the) Revolution Has Begun," *Wired* 2 (October 1994): 10.
29. Ric Manning, "Transferring Money by E-mail is Easy with PayPal, New Bank Services," *The Courier-Journal (Louisville, KY)* (March 18, 2000).
30. Katie Hafner, "Will That Be Cash or Cell Phone? Wireless Payment Systems Might Mean Dialing Into Your Own Wallet," *New York Times* (March 2, 2000).
31. Louis Rossetto, "The Original WIRED Manifesto," *Wired* (1993), https://www.wired.com/story/original-wired-manifesto/.
32. Gary Wolf, *Wired: A Romance* (New York: Random House, 2003), 12–14.
33. National Telecommunications and Information Administration, "Falling through the Net II: New Data on the Digital Divide," (July 28, 1998), https://www.ntia.gov/report/1998/falling-through-net-ii-new-data-digital-divide.
34. Malecki and Hu, "A Wired World: The Evolving Geography of Submarine Cables and the Shift to Asia," 370.
35. Daniel Pimienta, Daniel Prado, and Álvaro Blanco, *Twelve Years of Measuring Linguistic Diversity in the Internet: Balance and Perspectives* (Paris: UNESCO, 2009), 40.

36. Schiller, *Digital Capitalism: Networking the Global Market System*, 35.
37. Stephanie Ricker Schulte, "Technology and Networks of Communication," in *The Cambridge History of America and the World*, eds. David Engerman, Max Paul Friedman, and Melani McAlister (Cambridge, UK: Cambridge University Press, 2021), 617.
38. Schiller, *Digital Capitalism: Networking the Global Market System*, xiv.
39. Friedrich v. Gottl-Ottlilienfeld, *Fordismus? Von Frederick W. Taylor zu Henry Ford* (Jena: Gustav Fischer, 1925), 35. Thanks to Stefan Link for a scan of the original text.
40. David Kleinbard, "The $1.7 Trillion Dot.com Lesson," *CNNMoney* (November 9, 2000).
41. Peter Thiel and Blake Masters, *Zero to One: Notes on Startups, or How to Build the Future* (New York: Crown Business, 2014), 170.
42. Ibid., 21.
43. Ibid., 34.
44. Ibid., 24.
45. Ibid., 33.
46. Ibid., 83.
47. Ibid., 188.
48. Ibid.
49. Harris Fricker, Musk's X.com cofounder, quoted in Vance, *Elon Musk: Tesla, SpaceX, and the Quest for a Fantastic Future*, 83.
50. Thiel and Masters, *Zero to One*, 81.
51. Davidson and Rees-Mogg, *The Sovereign Individual*. Thiel wrote the preface for the twenty-fifth anniversary reissue. James Dale Davidson and William Rees-Mogg, *The Sovereign Individual: Mastering the Transition to the Information Age* (New York: Touchstone, 2020).
52. Peter Thiel, "The Education of a Libertarian," *Cato Unbound* (13 Apr 2009), https://www.cato-unbound.org/2009/04/13/peter-thiel/education-libertarian.
53. "1998: Elon Musk on His Early Silicon Valley days, Future of the Internet," *CBS Sunday Morning*, https://www.youtube.com/watch?v=zfwK5BvZrY4.
54. Marc Andreessen, "Why Software is Eating the World," *Andreessen Horowitz* (August 20, 2011), https://a16z.com/why-software-is-eating-the-world/.

3. SOVEREIGNTY AS A SERVICE

1. For Rumsfeld speech, see https://www.youtube.com/watch?v=MfMjdKElgqY. See also the description in Naomi Klein, *The Shock Doctrine: The Rise of Disaster Capitalism* (New York: Metropolitan Books, 2007), 360–63.
2. Donald Rumsfeld, "Transforming the Military," *Foreign Affairs* (May/June 2002).
3. Robert D. Kaplan, "What Rumsfeld Got Right," *The Atlantic* (July/August 2008), https://www.theatlantic.com/magazine/archive/2008/07/what-rumsfeld-got-right/306870/; Philip Gold, "Rumsfeld's Revolution," *Discovery Institute* (June 30, 2001), https://www.discovery.org/a/655/.
4. Arthur K. Cebrowski and John W. Raymond, "Operationally Responsive Space: A New Defense Business Model," *The US Army War College Quarterly: Parameters* 35, no. 2 (2005): x–xii, 8; Peter Dombrowski and Eugene Gholz, *Buying Military Transformation: Technological Innovation and the Defense Industry* (New York: Columbia University Press, 2006).
5. Bill Keller, "The Fighting Next Time," *New York Times* (March 10, 2002), https://www.nytimes.com/2002/03/10/magazine/the-fighting-next-time.html.
6. Wilfred P. Deac, "The Navy's Spy Missions in Space," *U.S. Naval Institute* (April 1, 2008), https://www.usni.org/magazines/naval-history-magazine/2008/april/navys-spy-missions-space.
7. John A. Tirpak, "Precision: The Next Generation," *Air Force Magazine* (November 2003), https://www.airandspaceforces.com/PDF/MagazineArchive/Documents/2003/November%202003/1103precision.pdf.
8. Rudi Williams, "Cold War Space Approach Must Change," *American Forces Press Service* (April 2, 2004), https://www.af.mil/News/Article-Display/Article/137254/cold-war-space-approach-must-change.
9. M. Hurley et al., "Engineering a Responsive, Low Cost, Tactical Satellite, TacSat-1," *Proceedings of the AIAA/USU Conference on Small Satellites, Logan, UT* (Aug 9–12, 2004). Also "TacSat-1 (Tactical

Satellite-1)" entry in European Space Agency's eoPortal, https://www.eoportal.org/satellite-missions/tacsat-1#eop-quick-facts-section, and Lt. Col. Jay Ramond et al., "TacSat-1 and a Path to Tactical Space," 2nd Responsive Space Conference (April 19–22, 2004).

10. Jeremy Singer, "Military Transformation Pioneer Arthur Cebrowski Dies at 63," *SpaceNews* (November 21, 2005), https://spacenews.com/military-transformation-pioneer-arthur-cebrowski-dies-63/.

11. Jeremy Singer, "Responsive Space," *Air Force Magazine* (March 1, 2006); Cebrowski and Raymond, "Operationally Responsive Space: A New Defense Business Model."

12. Vance, *Elon Musk: Tesla, SpaceX, and the Quest for a Fantastic Future*, 127–28; Isaacson, *Elon Musk*, 121. SpaceX never ended up launching the TacSat-1 satellite. After numerous delays, the Pentagon cancelled the TacSat-1 launch in 2007. See SpaceNews Editor, "Pentagon Cancels TacSat-1 Mission," *SpaceNews* (September 6, 2007), https://spacenews.com/pentagon-cancels-tacsat-1-mission/.

13. Elon Musk interview on CNN (2001), https://www.youtube.com/watch?v=l2mLcdb7cEU.

14. Isaacson, *Elon Musk*, 92.

15. Vance, *Elon Musk: Tesla, SpaceX, and the Quest for a Fantastic Future*, 89.

16. One of Musk's first public discussions of his desire to make humanity "multiplanetary" appeared in his remarks at the National Space Society's 24th annual International Space Development Conference, (May 19–22, 2005), Washington DC, https://www.youtube.com/watch?v=8vBqtKQx7jg.

17. "Elon Musk Captured by Rainn Wilson! | Metaphysical Milkshake."

18. "Elon Musk's First Public Speech – Talks Paypal and SpaceX 2003 [Entrepreneurial Thought Leaders Lecture at Stanford University]," (October 8, 2003), https://www.youtube.com/watch?v=n3yfaoMUo1s.

19. Elon Musk interview on CNN (2004), https://www.youtube.com/watch?v=ao5OdiwKp5k. See also Tim Fernholz, "What Happens When the US Stops Funding the Science Behind SpaceX?" Bloomberg (October 24, 2025), https://www.bloomberg.

com/news/articles/2025-10-24/as-trump-defunds-nasa-elon-musk-s-spacex-runs-on-borrowed-science.
20. James F. Feltz and Ralph Vartabedian, "California Largely Unscathed in Lockheed Cuts," *Los Angeles Times* (June 27, 1995).
21. "Military expenditure (% of GDP)—United States," World Bank Open Data, https://data.worldbank.org.
22. "Elon Musk's First Public Speech – talks Paypal and SpaceX 2003 [Entrepreneurial Thought Leaders Lecture at Stanford University]."
23. Max Chafkin, *The Contrarian: Peter Thiel and Silicon Valley's Pursuit of Power* (New York: Penguin Press, 2021), 113; Jeffrey Rosen, "Silicon Valley's Spy Game," *New York Times Magazine* (April 14, 2002), https://www.nytimes.com/2002/04/14/magazine/silicon-valley-s-spy-game.html.
24. Klein, *The Shock Doctrine: The Rise of Disaster Capitalism*, 358-60.
25. Congressional Research Service, "Department of Defense Contractor and Troop Levels in Afghanistan and Iraq: 2007–2020," (February 22, 2021), https://sgp.fas.org/crs/natsec/R44116.pdf.
26. David Johnston and John M. Broder, "F.B.I. Says Guards Killed 14 Iraqis Without Cause," *The New York Times* (November 14, 2007), https://www.nytimes.com/2007/11/14/world/middleeast/14blackwater.html.
27. Spencer Ackerman, "Snowden Leak Shines Light on US Intelligence Agencies' Use of Contractors," *Guardian* (June 10, 2013), https://www.theguardian.com/world/2013/jun/10/edward-snowden-booz-allen-hamilton-contractors.
28. The contract was awarded as part of FALCON, a joint project between DARPA and the Air Force—not to be confused with SpaceX's Falcon 1 rocket. Jan Walker, "DARPA, Air Force Kick Off Falcon Phase II Small Launch Vehicle," *SpaceNews* (September 17, 2004), https://spacenews.com/darpa-air-force-kick-off-falcon-phase-ii-small-launch-vehicle/.
29. "SpaceX Awarded $100 Million Contract From U.S. Air Force for Falcon I," *SpaceNews* (May 2, 2005), https://spacenews.com/spacex-awarded-100-million-contract-from-us-air-force-for-falcon-i/.
30. Jack Kuhr, "2024 Orbital Launch Attempts by Country," *Payload*, 2025, https://payloadspace.

com/2024-orbital-launch-attempts-by-country/; Todd Harrison, "Space Trends in 2024," *American Enterprise Institute* (January 13, 2025), https://www.aei.org/op-eds/space-trends-in-2024/; Jeff Foust, "SpaceX launch surge helps set new global launch record in 2024," *Spacenews* (January 1, 2025), https://spacenews.com/spacex-launch-surge-helps-set-new-global-launch-record-in-2024/.

31. Vance, *Elon Musk: Tesla, SpaceX, and the Quest for a Fantastic Future*, 112–13. In 2008, SpaceX relocated its headquarters from El Segundo to the neighboring municipality of Hawthorne. In 2024, SpaceX moved its headquarters once again, this time to South Texas.
32. djvlad, "Errol Musk, Father of Elon Musk, Tells His Life Story (Full Interview)."
33. California Employment Development Department, "Employment by Industry Data," https://labormarketinfo.edd.ca.gov/data/employment-by-industry.html.
34. "Musk's Remarks at the National Space Society's 24th Annual International Space Development Conference."
35. Ibid.
36. Kevin Brogan, quoted in Vance, *Elon Musk: Tesla, SpaceX, and the Quest for a Fantastic Future*, 188.
37. The company was Barber–Nichols, and the executives were Robert Linden, Gary Frey and Mike Forsha; quoted in Vance, *Elon Musk: Tesla, SpaceX, and the Quest for a Fantastic Future*, 197.
38. Isaacson, *Elon Musk*, 448.
39. Miriam Posner, "Agile and the Long Crisis of Software," *Logic* (March 27, 2022), https://logicmag.io/clouds/agile-and-the-long-crisis-of-software/.
40. Quoted in Vance, *Elon Musk: Tesla, SpaceX, and the Quest for a Fantastic Future*, 141.
41. Ibid., 114.
42. Andrew Chaikin, "Is SpaceX Changing the Rocket Equation?," *Air & Space Magazine* (January 2012), https://www.smithsonianmag.com/air-space-magazine/is-spacex-changing-the-rocket-equation-132285884/.
43. James P. Womack et al., *The Machine that Changed the World* (New York: Rawson Associates, 1990).

44. "Manifesto for Agile Software Development" (2001), https://agilemanifesto.org/; Mary Poppendieck et al, *Lean Software Development: An Agile Toolkit* (Boston: Addison-Wesley Professional, 2003).
45. Isaacson, *Elon Musk*, 227, 72, 84–5.
46. Bruce Betts and Mat Kaplan, "A Conversation with Elon Musk of SpaceX," *Planetary Radio* (February 16, 2009), https://www.planetary.org/planetary-radio/328; Isaacson, *Elon Musk*, 112.
47. "Musk's Remarks at the National Space Society's 24th Annual International Space Development Conference."
48. Isaacson, *Elon Musk*, 284.
49. Isaacson, *Elon Musk*, 113.
50. "Elon Musk's First Public Speech – Talks Paypal and SpaceX 2003 [Entrepreneurial Thought Leaders Lecture at Stanford University]."
51. Tim Fernholz, *Rocket Billionaires: Elon Musk, Jeff Bezos, and the New Space Race* (Houghton Mifflin Harcourt, 2020), 84.
52. Betts and Kaplan, "A Conversation with Elon Musk of SpaceX."
53. Isaacson, *Elon Musk*, 109–10.
54. Ibid., 157.
55. Vance, *Elon Musk: Tesla, SpaceX, and the Quest for a Fantastic Future*, 141. Fernholz, *Rocket Billionaires*, 83; Ashlee Vance, *When the Heavens Went on Sale: The Misfits and Geniuses Racing to Put Space Within Reach* (New York, NY: Ecco, 2023), 38.
56. Fernholz, *Rocket Billionaires*, 82–3; James Gattuso, "Brilliant Pebbles: The Revolutionary Idea for Strategic Defense," *The Heritage Foundation*, https://www.heritage.org/defense/report/brilliant-pebbles-the-revolutionary-idea-strategic-defense.
57. Vance, *When the Heavens Went on Sale*, 47.
58. Quoted in ibid., 48.
59. Mario Pianta, *New Technologies Across the Atlantic: US Leadership or European Autonomy?* (Hemel Hempstead, UK: Harvester/Wheatshead, 1988).
60. Fernholz, *Rocket Billionaires*, 99–101; Vance, *When the Heavens Went on Sale*, 43.

61. Isaacson, *Elon Musk*, 101; Eric Berger, *Liftoff: Elon Musk and the Desperate Early Days that Launched SpaceX* (New York, NY: William Morrow, 2021), 13.
62. Musk's remarks at Stanford University (October 7, 2015), https://www.youtube.com/watch?v=SVk1hb0ZOrE.
63. Jason Vest, "Darth Rumsfeld," *The American Prospect* (December 19, 2001), https://prospect.org/api/content/465681f4-bf07-5bde-921c-8cdd606fd451/.
64. "NASA Presolicitation Notice: Kistler K-1 Pre-Flight and Post-flight Data," *SpaceRef* (February 3, 2004), https://spaceref.com/status-report/nasa-presolicitation-notice-kistler-k-1-pre-flight-and-post-flight-data/. Fernholz, *Rocket Billionaires*, 115.
65. Isaacson, *Elon Musk*, 122; Berger, *Liftoff*, 108–10.
66. "Prepared Statement by Elon Musk at a Senate Hearing on Space Shuttle and the Future of Space Launch Vehicles," *SpaceRef* (May 5, 2004), https://spaceref.com/status-report/prepared-statement-by-elon-musk-at-a-senate-hearing-on-space-shuttle-and-the-future-of-space-launch-vehicles/.
67. "NASA's Griffin Sees Commercial Space Trips," *NBC News* (November 16, 2005), https://www.nbcnews.com/id/wbna10066313.
68. NASA, *Commercial Orbital Transportation Services: A New Era in Spaceflight* (Washington, DC: Government Printing Office, 2014); Berger, *Liftoff*, 110.
69. Brian Berger, "SpaceX, Rocketplane Kistler Win NASA COTS Competition," *Space.com* (August 18, 2006), https://www.space.com/2768-spacex-rocketplane-kistler-win-nasa-cots-competition.html.
70. Fernholz, *Rocket Billionaires*, 130.
71. Quoted in Fernholz, *Rocket Billionaires*, 131.
72. NASA, *Commercial Orbital Transportation Services: A New Era in Spaceflight*. Isaacson, *Elon Musk*, 123.
73. Graham Bensinger, "Kimbal Musk: Working with Elon, Taking Risks on Tesla, and Building The Kitchen," *In Depth with Graham Bensinger* (March 29, 2025), https://www.youtube.com/watch?v=KDUZrVL-cEhY. Musk's remarks at the Churchill Club, April 8, 2009, https://www.youtube.com/watch?v=n1joyHOxcL0. NASA,

Commercial Orbital Transportation Services: A New Era in Spaceflight.
74. "Elon Musk's First Public Speech – Talks Paypal and SpaceX 2003 [Entrepreneurial Thought Leaders Lecture at Stanford University]."
75. Vance, *Elon Musk: Tesla, SpaceX, and the Quest for a Fantastic Future*, 241–2.
76. Quoted in ibid., 203.
77. Berger, *Liftoff*, 53–4.
78. Jonathan McDowell, "Starlink Statistics," https://planet4589.org/space/con/star/stats.html; Petroc Taylor, "Active satellites by category/operator," *Statista* (September 3, 2025), https://www.statista.com/statistics/1224643/active-satellite-by-operator/; Anthony Cuthbertson, "Elon Musk is Taking Control of Space and the Internet – It Could End Badly," *The Independent* (October 25, 2025), https://www.the-independent.com/tech/elon-musk-space-internet-starlink-b2851503.html.
79. Joe Supan, "Musk's Starlink Internet Is Now Available in Over 100 Countries," *CNET* (February 18, 2025), https://www.cnet.com/home/internet/starlink-internet-is-available-in-over-100-countries/.
80. Michael Sheetz, "SpaceX's Starlink Wins Nearly $900 Million in FCC Subsidies to Bring Internet to Rural Areas," *CNBC* (December 7, 2020), https://www.cnbc.com/2020/12/07/spacex-starlink-wins-nearly-900-million-in-fcc-subsidies-auction.html; Cecilia Kang, "Federal Grant Program Opens Door to Elon Musk's Starlink," *New York Times* (March 6, 2025), https://www.nytimes.com/2025/03/05/technology/broadband-rules-elon-musk-starlink.html; Tamara Chuang, "Starlink, Amazon Asking for $300M of Colorado's Broadband Money after Federal Rule Changes," *The Colorado Sun* (July 25, 2025), https://coloradosun.com/2025/07/25/cheaper-wireless-satellite-internet-trumps-fiber-colorado-broadband-bead/.
81. Joey Roulette and Marisa Taylor, "Exclusive: Musk's SpaceX is Building Spy Satellite Network for US Intelligence Agency, Sources Say," Reuters (March 16, 2024), https://www.reuters.com/

technology/space/musks-spacex-is-building-spy-satellite-network-us-intelligence-agency-sources-2024-03-16/.
82. Wang Peiwen, Zhang Huang and Zhang Kaiyue, "Starlink Militarization: Challenges and Responses to Space Intelligence and Information Security," *CSIS Interpret: China* (January 9, 2024), https://interpret.csis.org/translations/starlink-militarization-challenges-and-responses-to-space-intelligence-and-information-security/.
83. Isaacson, *Elon Musk*, 428–34.
84. Ronan Farrow, "Elon Musk's Shadow Rule," *The New Yorker* (August 21, 2023), https://www.newyorker.com/magazine/2023/08/28/elon-musks-shadow-rule.
85. https://x.com/elonmusk/status/1699913329261813809, September 7, 2023.
86. https://x.com/WalterIsaacson/status/1700342242290901361, September 8, 2023; https://x.com/WalterIsaacson/status/1700522506363248665, September 9, 2023.
87. Joey Roulette, Cassell Bryan-Low and Tom Balmforth, "Musk Ordered Shutdown of Starlink Satellite Service as Ukraine Retook Territory from Russia," Reuters (July 25, 2025), https://www.reuters.com/investigations/musk-ordered-shutdown-starlink-satellite-service-ukraine-retook-territory-russia-2025-07-25/.
88. Rafe Uddin and Stephen Morris, "Starlink's Rapid Global Rollout Complicated by Elon Musk's Ties to Donald Trump," *Financial Times* (March 23, 2025).
89. Teresa Guerrero, "Europe's Most Advanced Secure Communications Satellite is Spanish and Will be Launched Tonight with a Rocket from Elon Musk: 'It will give us strategic sovereignty,'" *El Mundo America* (January 29, 2025), https://www.mundoamerica.com/news/2025/01/29/6799f3c3fdddff84968b45d8.html.

4. ELECTRIC AUTONOMY

1. "Obama's Speech in Lansing, Michigan," *New York Times* (August 4, 2008), https://www.nytimes.com/2008/08/04/us/politics/04text-obama.html.

2. Thomas L. Friedman, "The Power of Green," *New York Times Magazine* (April 15, 2007), https://www.nytimes.com/2007/04/15/opinion/15iht-web-0415edgreen-full.5291830.html.
3. "Tesla 2024 Shareholder Meeting," https://www.youtube.com/watch?v=remZ1KMR_Z4.
4. Cory Doctorow, *Enshittification: Why Everything Suddenly Got Worse and What to Do About It* (New York: MCD, 2025).
5. Lee Brodie, "Cramer: Does Your Portfolio Have FANGs?," *CNBC* (February 5, 2013), https://www.cnbc.com/2013/02/05/cramer-does-your-portfolio-have-fangs.html.
6. Molly McHugh, "Tesla's Cars Now Drive Themselves, Kinda," *Wired* (October 14, 2015), https://www.wired.com/2015/10/tesla-self-driving-over-air-update-live.
7. For the full narrative, see Tim Higgins, *Power Play: Tesla, Elon Musk, and the Bet of the Century* (New York: Doubleday, 2021), chapters 2-8.
8. Margaret Pugh O'Mara, *The Code: Silicon Valley and the Remaking of America* (New York: Penguin Press, 2019), 395–6.
9. John Doerr, "Salvation (and profit) in greentech" (March 2007), https://www.ted.com/talks/john_doerr_salvation_and_profit_in_greentech.
10. Devashree Saha and Mark Muro, "Cleantech Venture Capital: Continued Declines and Narrow Geography Limit Prospects," *Brookings Institution* (May 16, 2017), https://www.brookings.edu/articles/cleantech-venture-capital-continued-declines-and-narrow-geography-limit-prospects/.
11. Quoted in Adam Lashinsky and Marc Gunther, "Cleanup Crew," *Fortune* (February 12, 2008), https://fortune.com/2008/02/12/al-gore-kleiner/.
12. These were the Energy Policy Act of 2005 and Energy Independence and Security Act of 2007.
13. "Conversation with Elon Musk (Tesla Motors) – Web 2.0 Summit 08," (November 10, 2008), https://www.youtube.com/watch?v=gVwmNaPsxLc.
14. Vance, *Elon Musk: Tesla, SpaceX, and the Quest for a Fantastic Future*, 285–8.

15. Isaacson, *Elon Musk*, 179, 193.
16. Committee on Oversight and Government Reform, "The Department of Energy's Disastrous Management of Loan Guarantee Programs," (March 20, 2012): 48.
17. Randall Stross, "Only the Rich Can Afford It. Should Taxpayers Back It?" *New York Times* (October 30, 2008).
18. Diarmuid O'Connell, "Clearing the Air on Our DOE Loan" (September 28, 2009), *Tesla Motors*, https://web.archive.org/web/20130121230258/http://www.teslamotors.com/blog/clearing-air-our-doe-loan.
19. Quoted in Claire Cain Miller, "An All-Electric Sedan, Awaiting Federal Aid," *New York Times* (March 26, 2009).
20. Corbin Hair, "Tesla Built Musk's Vast Wealth Through Climate Credits. Trump May End Them," *Politico* (January 18, 2025), https://www.politico.com/news/2025/01/18/musk-tesla-climate-credits-trump-00198794; Edward Niedermeyer, *Ludicrous: the unvarnished story of Tesla Motors* (Dallas, TX: BenBella Books, 2019), 72–3, 117–18.
21. "Robert Lutz; Electric Cars," *Charlie Rose Show* (November 9, 2011), https://charlierose.com/videos/23446.
22. Steven Mufson, "As Fuel Prices Fall, Will Push For Alternatives Lose Steam?," *Washington Post* (October 20, 2008).
23. Matthew DeBord, "Tesla Bought an Old GM-Toyota Factory and Made it Cool—But In Its Former Life it Built a Lot More Cars," *Business Insider* (October 27, 2017), https://www.businessinsider.com/tesla-factory-built-more-cars-when-it-belonged-gm-and-toyota-2017-10.
24. Isaacson, *Elon Musk*, 219.
25. Joan Fitzgerald, "Solar Eclipse? Can the U.S. Have a Coherent Solar Policy in the Face of China's Strategic Trade Moves?" *The American Prospect* (Summer 2016), https://link.gale.com/apps/doc/A459510551/AONE?u=mlin_b_bumml&sid=bookmark-AONE&xid=4ca83078.
26. "Oil Production in the United States from 1998 to 2024," *Statista* (July 2, 2025), https://www.statista.com/statistics/265215/us-oil-production-in-million-metric-tons; "Natural Gas Production in the United States from 1998 to 2024," *Statista*

(July 8, 2025), https://www.statista.com/statistics/265331/natural-gas-production-in-the-us/.
27. Benjamin Gaddy et al, "Venture Capital and Cleantech: The Wrong Model for Clean Energy Innovation," *MIT Energy Initiative* (July 2016), https://energy.mit.edu/wp-content/uploads/2016/07/MITEI-WP-2016-06.pdf.
28. Mazzucato, *The Entrepreneurial State*.
29. Nathan Donato-Weinstein, "Exclusive: Tesla Snaps Up Former Solyndra Building in Huge Fremont Expansion," *Silicon Valley Business Journal* (June 11, 2015), https://www.bizjournals.com/sanjose/news/2015/06/11/exclusive-tesla-snaps-up-former-solyndra-building.html.
30. "Elon Musk," *Charlie Rose Show* (August 11, 2009), https://charlierose.com/videos/12550.
31. Tesla Motors Inc., "Registration Statement on Form S-1" (January 29, 2010), https://www.sec.gov/Archives/edgar/data/1318605/000119312510017054/ds1.htm.
32. John Reed, "Electric Dreams," *Financial Times* (January 8, 2008).
33. See annual World Trade Summary data on https://wits.worldbank.org/.
34. Higgins, *Power Play*, 188.
35. Hamish McKenzie, *Insane Mode: How Elon Musk's Tesla Sparked an Electric Revolution to End the Age of Oil* (New York: Dutton, 2018), 191.
36. Richard Waters, "Tesla to Gamble on Nevada Battery Plant," *Financial Times* (September 4, 2014).
37. Andy Sharman and Richard Waters, "Time to Accelerate," *Financial Times* (September 11, 2015).
38. "CATL Reveals 2020 Installation Figures, Tightly Aligned with Adamas Tracker Data," *Adamas Inside* (April 29, 2021), https://www.adamasintel.com/catl-reveals-2020-installation-figures/; Tom Hancock, Leo Lewis, and Henry Sanderson, "Battery Power," *Financial Times* (March 6, 2017).
39. Pilita Clark, "The Big Green Bang," *Financial Times* (May 19, 2017). Jonathan E. Hillman, *The Digital Silk Road: China's Quest to Wire the World and Win the Future* (New York: Harper Business, 2021); Leslie

Hook and Henry Sanderson, "The New Green Order," *Financial Times* (February 6, 2021).

40. Anthea Roberts, Henrique Choer Moraes, and Victor Ferguson, "Toward a Geoeconomic Order," *Journal of International Economic Law* 22, no. 4 (Dec 2019), 655–76.
41. Ryan McMorrow, "Tesla Lines Up $1.6bn Financing for Model 3 Plant in Shanghai," *Financial Times* (December 28, 2019).
42. Dana Hull, "Musk Tells Tesla Shareholders Consumer Demand Not a Problem," Bloomberg (June 11, 2019), https://www.bloomberg.com/news/articles/2019-06-12/musk-tells-tesla-shareholders-consumer-demand-not-a-problem?sref=apOkUyd1.
43. Richard Milne and Ben Hall, "Powering Up the Industrial Base," *Financial Times* (December 3, 2019).
44. Andreas Malm and Wim Carton, *Overshoot* (New York: Verso, 2024).
45. Mark Matousek, "Tesla Just Announced a Giant New Battery," *Business Insider* (July 30, 2019), https://www.businessinsider.com/tesla-announces-megapack-giant-new-battery-product-2019-7; "Elon Musk Says He Can Rebuild Puerto Rico's Power Grid with Solar," *BBC* (October 6, 2017), https://www.bbc.com/news/world-us-canada-41524220.
46. "Elon Musk Says He Can Rebuild Puerto Rico's Power Grid with Solar."
47. "Tesla Hits Record on Hopes for Battery Pack 'Gigafactory'," *Financial Times* (February 26, 2014).
48. Kirsten Korosec, "Tesla Has a New Energy Product Called Megapack," *TechCrunch* (July 29, 2019), https://techcrunch.com/2019/07/29/tesla-has-a-new-energy-product-called-megapack/.
49. "Elon Musk at Tesla, Inc. 2021 Annual Meeting of Stockholders," (October 7, 2021), https://www.youtube.com/watch?v=bH51-loeLgM.
50. Robert Ferris, "Why Tesla's Turn to Robots has Divided Wall Street," *CNBC* (April 16, 2025).
51. Danielle Muoio, "Elon Musk: Tesla's Factory Will Be an 'Alien Dreadnought' by 2018," *Business Insider* (October 27, 2016), https://www.businessinsider.com/elon-musk-tesla-factory-alien-dreadnought-2016-10; Matthew DeBord, "Elon Musk's Big Battery

Plans Include Another Shot at His 'Alien Dreadnought' Factory Dream," *Business Insider* (September 23, 2020), https://www.businessinsider.com/elon-musk-revives-his-alien-dreadnought-factory-dream-for-batteries-2020-9.

52. Rem Koolhaas, *Delirious New York: A Retroactive Manifesto for Manhattan* (New York: Oxford University Press, 1978); Chuihua Judy Chung et al., *Project on the City I: Great Leap Forward* (Cologne: Taschen, 2001).

53. "Countryside, The Future," *Solomon R. Guggenheim Museum* (2020): 5-6, https://www.guggenheim.org/wp-content/uploads/2020/02/Countryside-The-Future_PressKit_021920.pdf.

5. ATTENTION ALCHEMY

1. Richard MacManus, "The First Web 2.0 Conference in 2004: A New Bubble Begins," *Cybercultural* (November 8, 2023), https://cybercultural.com/p/003-the-first-web-20-conference-2004/; Tim O'Reilly, "Ask Jeff Bezos, Adam Bosworth, John Doerr, Eddy Cue ...," *O'Reilly Developer Weblogs* (September 21, 2004), https://web.archive.org/web/20041009150551/http://www.oreillynet.com/pub/wlg/5630.

2. Tim O'Reilly, "What Is Web 2.0: Design Patterns and Business Models for the Next Generation of Software," *O'Reilly* (September 30, 2005), https://www.oreilly.com/pub/a/web2/archive/what-is-web-20.html.

3. Quoted in MacManus, "The First Web 2.0 Conference in 2004: A New Bubble Begins."

4. O'Reilly, "What Is Web 2.0: Design Patterns and Business Models for the Next Generation of Software."

5. Shoshana Zuboff, *The Age of Surveillance Capitalism* (New York: PublicAffairs, 2019).

6. Nick Srnicek, *Platform Capitalism* (Cambridge, UK: Polity, 2017), 30.

7. Pew Research Center, "Newspapers Fact Sheet," November 10, 2023, https://www.pewresearch.org/journalism/fact-sheet/newspapers/. IAB & PWC, "Internet Advertising Revenue Report: Full-year 2024 Results" (April 2025), 30–31.

8. Zuboff, *The Age of Surveillance Capitalism*, 87.
9. William Davies, "The Reaction Economy," *London Review of Books* (March 2, 2023), https://www.lrb.co.uk/the-paper/v45/n05/william-davies/the-reaction-economy.
10. William J. Clinton, "Remarks on Signing the North American Free Trade Agreement Implementation Act," *American Presidency Project* (December 8, 1993), https://www.presidency.ucsb.edu/documents/remarks-signing-the-north-american-free-trade-agreement-implementation-act.
11. Leonid Bershidsky, "There Will Be No More Facebook Revolutions," Bloomberg (October 18, 2019), https://www.bloomberg.com/opinion/articles/2019-10-18/post-facebook-revolutions-will-be-organized-on-apps-like-telegram?sref=apOkUyd1.
12. Mattathias Schwartz, "Pre-Occupied," *The New Yorker* (November 28, 2011), https://www.newyorker.com/magazine/2011/11/28/pre-occupied.
13. Peter Hapak, "The Protester: A Portfolio," *TIME* (December 14, 2011), https://time.com/3783718/person-of-the-year-2011-protesters-2/.
14. Mabrouka M'Barek, "Enough with the 'Jasmine Revolution' Narrative: Tunisians Demand Dignity," *Middle East Eye* (April 18, 2016), https://www.middleeasteye.net/opinion/enough-jasmine-revolution-narrative-tunisians-demand-dignity.
15. Nitasha Tiku, "Ground Control to Silicon Valley," *BuzzFeed News* (June 6, 2016), https://www.buzzfeednews.com/article/nitashatiku/ground-control-to-silicon-valley; "What is Twitter | Jack Dorsey, DeRay Mckesson | Code Conference 2016," *On with Kara Swisher* (August 14, 2016), https://www.youtube.com/watch?v=LUzl_g7ipC4; Kate Conger and Ryan Mac, *Character Limit: How Elon Musk Destroyed Twitter* (New York: Penguin Press, 2024), 362.
16. Aja Romano, "A History of 'Wokeness'," *Vox* (October 9, 2020), https://www.vox.com/culture/21437879/stay-woke-wokeness-history-origin-evolution-controversy.
17. His invocation of the term was greeted mostly with an eyeroll and suspicions of cynicism. See Matt Miller, "Behold: The Most

Painful T-Shirt on the Internet," *Esquire* (June 2, 2016), https://www.esquire.com/style/news/a45458/jack-dorsey-stay-woke-shirt/.

18. Richard Dawkins, *The Selfish Gene* (Oxford, UK: Oxford University Press, 1976), 206.
19. Limor Shifman, *Memes in Digital Culture* (Cambridge, MA: The MIT Press, 2014), 20.
20. Caitlin Dewey, "Absolutely Everything You Need to Know to Understand 4chan, the Internet's Own Bogeyman," *Washington Post* (September 25, 2014), https://www.washingtonpost.com/news/the-intersect/wp/2014/09/25/absolutely-everything-you-need-to-know-to-understand-4chan-the-internets-own-bogeyman/; Brian Raftery, "King of Cheez: The Internet's Meme Maestro Turns Junk Into Gold," *Wired* (January 25, 2010), https://www.wired.com/2010/01/mf-cheezking/.
21. Whitney Phillips, *This Is Why We Can't Have Nice Things: Mapping the Relationship between Online Trolling and Mainstream Culture* (Cambridge, MA: MIT Press, 2016).
22. Mattathias Schwartz, "The Trolls Among Us," *New York Times* (August 3, 2008), https://www.nytimes.com/2008/08/03/magazine/03trolls-t.html. On the broader phenomenon see E. Gabriella Coleman, *Hacker, hoaxer, whistleblower, spy: the many faces of Anonymous* (New York: Verso, 2014).
23. Jonathan Weisman, "The Nazi Tweets of 'Trump God Emperor,'" *New York Times* (May 26, 2016), https://www.nytimes.com/2016/05/29/opinion/sunday/the-nazi-tweets-of-trump-god-emperor.html.
24. Elizabeth Chan, "Donald Trump, Pepe the Frog, and White Supremacists: An Explainer" (September 12, 2016), https://www.hillaryclinton.com/feed/donald-trump-pepe-the-frog-and-white-supremacists-an-explainer/.
25. "Pepe the Frog Meme Branded a 'Hate Symbol'," *BBC News* (September 28, 2016), https://www.bbc.com/news/world-us-canada-37493165.
26. Philip Bump, "The Hashtag that Ultimately Overtook #BlackLivesMatter for Online Activism? #MAGA," *Washington Post*

(July 11, 2018), https://www.washingtonpost.com/news/politics/wp/2018/07/11/the-hashtag-that-ultimately-overtook-blacklivesmatter-for-online-activism-maga/.

27. Joseph Bernstein, "Bad News: Selling the Story of Disinformation," *Harper's* (September 2021), https://harpers.org/archive/2021/09/bad-news-selling-the-story-of-disinformation/.
28. Nicholas A. John, "Sharing and Social Media: The Decline of a Keyword?," *new media & society* 26, no. 4 (2024): 1891–908.
29. Emily Stewart, "Lawmakers Seem Confused About What Facebook Does—and How to Fix It," *Vox* (2018), https://www.vox.com/policy-and-politics/2018/4/10/17222062/mark-zuckerberg-testimony-graham-facebook-regulations.
30. Chelsea Butkowski and Frances Corry, "Social Media's Midlife Crisis? How Public Discourse Imagines Platform Futures," *Social Media + Society* (April–June 2025): 1–13.
31. Brooke Auxier, "64% of Americans Say Social Media Have a Mostly Negative Effect on the Way Things Are Going in the U.S. Today," *Pew Research Center* (October 15, 2020), https://www.pewresearch.org/short-reads/2020/10/15/64-of-americans-say-social-media-have-a-mostly-negative-effect-on-the-way-things-are-going-in-the-u-s-today/.
32. Joanna Kavenna, "Shoshana Zuboff: 'Surveillance capitalism is an assault on human autonomy,'" *Guardian* (October 4, 2019), https://www.theguardian.com/books/2019/oct/04/shoshana-zuboff-surveillance-capitalism-assault-human-automomy-digital-privacy.
33. Zuboff, *The Age of Surveillance Capitalism*, 521.
34. For Musk's Twitter usage from 2011 to 2021, see https://www.visualcapitalist.com/a-decade-of-elon-musks-tweets-visualized/. Davey Alba et al., "Elon Musk Is Now X's Biggest Promoter of Anti-Immigrant Conspiracies," Bloomberg (October 24, 2024), https://www.bloomberg.com/graphics/2024-musk-x-election-influence-immigration/.
35. Conger and Mac, *Character Limit*, 37; Vance, *Elon Musk: Tesla, SpaceX, and the Quest for a Fantastic Future*, 205.
36. Conger and Mac, *Character Limit*, 36.

37. Phil Jones, "Silicon Telepathy," *LRB Blog* (June 25, 2021), https://www.lrb.co.uk/blog/2021/june/silicon-telepathy.
38. https://x.com/elonmusk/status/1026872652290379776, August 7, 2018.
39. Clare Duffy, "Elon Musk Wins Lawsuit Over 'funding secured' Tweet," *CNN* (2023), https://www.cnn.com/2023/02/03/cars/musk-tesla-tweet-lawsuit-jury.
40. https://x.com/elonmusk/status/1256239815256797184, May 1, 2020; Jessica Bursztynsky, "Tesla Shares Tank after Elon Musk Tweets the Stock Price is 'too high'," *CNBC* (May 1, 2020), https://www.cnbc.com/2020/05/01/tesla-ceo-elon-musk-says-stock-price-is-too-high-shares-fall.html.
41. Ryan Browne, "Bitcoin Spikes 20% After Elon Musk Adds #bitcoin to his Twitter Bio," *CNBC* (January 29, 2021), https://www.cnbc.com/2021/01/29/bitcoin-spikes-20percent-after-elon-musk-adds-bitcoin-to-his-twitter-bio.html.
42. Marco D'Eramo, "Iron Musk," *Sidecar* (June 15, 2022), https://newleftreview.org/sidecar/posts/iron-musk.
43. Conger and Mac, *Character Limit*, 24–25.
44. Nick Bilton, "Elon Musk's Totally Awful, Batshit-Crazy, Completely Bonkers, Most Excellent Year," *Vanity Fair* (December 2020), https://www.vanityfair.com/news/2020/11/elon-musks-totally-awful-batshit-crazy-most-excellent-year.
45. Brian Schwartz and Lora Kolodny, "Elon Musk Said He Prefers to Stay Out of Politics – His Lobbying Efforts, Campaign Donations and Tweets Say Otherwise," *CNBC* (November 9, 2021), https://www.cnbc.com/2021/11/09/elon-musk-tesla-spacex-spend-millions-to-influence-politics-and-policy.html.
46. Nick Srnicek, "The New Conglomerates," *Platforms & Society* 1 (2024): 1-2.
47. Cédric Durand, *How Silicon Valley Unleashed Techno-Feudalism: The Making of the Digital Economy* (New York: Verso, 2024), 37.
48. PewDiePie, "Will Smith hosts Meme Review w/ Elon Musk," (February 22, 2019), https://www.youtube.com/watch?v=zpWYQ1YtgnI.
49. https://x.com/elonmusk/status/1153155448012300288, July 22, 2019.

50. James Benedict, Rebecca Elliott, and Inti Pacheco, "Twitter, Tesla and Copious Emojis: What and When Elon Musk Tweets," *Wall Street Journal* (April 29, 2022), https://www.wsj.com/tech/elon-musks-twitter-feed-sheds-light-on-twitters-prospective-owner-11651234142.
51. Bilton, "Elon Musk's Totally Awful, Batshit-Crazy, Completely Bonkers, Most Excellent Year."
52. Ryan M. Milner, *The World Made Meme: Public Conversations and Participatory Media* (Cambridge, MA: The MIT Press, 2016), 79. See also https://knowyourmeme.com/memes/doge.
53. Phillips, *This Is Why We Can't Have Nice Things*, 138.
54. Original Dogecoin website, accessible at https://web.archive.org/web/20131207010921/http://dogecoin.com/.
55. "Fuck Yeah Dogecoin," https://fuckyeahdogecoin.tumblr.com/post/74135150972/scruppy-doge-coin-its-happening.
56. https://x.com/elonmusk/status/1113009339743100929, April 2, 2019.
57. https://x.com/elonmusk/status/1357241340313141249, February 4, 2021.
58. Matt Levine, "Elon Musk Had Fun with Dogecoin," Bloomberg (June 5, 2023), https://www.bloomberg.com/opinion/articles/2023-06-05/elon-musk-had-fun-with-dogecoin?sref=apOkUyd1.
59. https://x.com/elonmusk/status/1531699416490557440 (May 31, 2022).
60. Levine, "Elon Musk Had Fun with Dogecoin."
61. Sabrina Toppa, "Elon Musk's Dogecoin Insider Trading Case Dismissed," *Yahoo Finance* (August 30, 2024), https://www.thestreet.com/crypto/markets/elon-musks-dogecoin-insider-trading-case-dismissed.
62. Eric Milstein and David Wessel, "What Did the Fed Do in Response to the COVID-19 Crisis?," *Brookings Institution* (January 2, 2024), https://www.brookings.edu/articles/fed-response-to-covid19/.
63. Jeff Kauflin, "The Inside Story of Robinhood's Billionaire Founders, Option Kid Cowboys and the Wall Street Sharks that Feed on Them," *Forbes* (August 19, 2020), https://www.forbes.com/sites/jeffkauflin/2020/08/19/

the-inside-story-of-robinhoods-billionaire-founders-option-kid-cowboys-and-the-wall-street-sharks-that-feed-on-them/.
64. https://x.com/elonmusk/status/1354174279894642703, January 26, 2021.
65. https://x.com/elonmusk/status/1357269755112148993, February 4, 2021.
66. Quoted in Cam Wilson, "Australian Dogecoin Creator Jackson Palmer on Grifts, Elon Musk, Crypto Bubbles and Pauline Hanson," *Crikey* (May 30, 2022), https://www.crikey.com.au/2022/05/30/dogecoin-jackson-palmer-elon-musk-crypto-bubble-pauline-hanson/.
67. https://x.com/elonmusk/status/1276418907968925696 (June 26, 2020).
68. Matt Levine, "Elon Musk's Bitcoin Fun Continues," Bloomberg (May 17, 2021), https://www.bloomberg.com/opinion/articles/2021-05-17/elon-musk-controls-bitcoin-and-dogecoin-prices-with-pure-magic?sref=apOkUyd1.
69. Levine, "Elon Musk Had Fun with Dogecoin."
70. Charlie Warzel, "Elon Musk Looks Desperate," *The Atlantic* (March 12, 2025), https://www.theatlantic.com/technology/archive/2025/03/elon-musk-human-meme-stock/682023/.

6. CYBERNETIC COLLECTIVES

1. Alex Hanna et al., "Lines of Sight," *Logic* (December 20, 2020), https://logicmag.io/commons/lines-of-sight/.
2. Fred Pope, "Tesla's Neural Network Revolution: How Full Self-Driving Replaced 300,000 Lines of Code with AI," (June 24, 2025), https://www.fredpope.com/blog/machine-learning/tesla-fsd-12.
3. Isaacson, *Elon Musk*, 242.
4. Nick Bostrom, *Superintelligence: Paths, Dangers, Strategies* (Oxford: Oxford University Press, 2014).
5. Isaacson, *Elon Musk*, 241.
6. Karen Hao, *Empire of AI: Dreams and Nightmares in Sam Altman's OpenAI* (New York: Penguin Press, 2025), 24.

7. Emily Price, "Elon Musk Nominated For 'Luddite' of the Year Prize Over Artificial Intelligence Fears," *Guardian* (December 23, 2015), https://www.theguardian.com/technology/2015/dec/24/elon-musk-nominated-for-luddite-of-the-year-prize-over-artificial-intelligence-fears.
8. "Elon Musk : How to Build the Future," *Y Combinator Startup School* (September 5, 2016), https://www.youtube.com/watch?v=tnBQmEqBCY0.
9. Quoted in Kelsey Piper, "Why Elon Musk Fears Artificial Intelligence," *Vox* (November 2, 2018), https://www.vox.com/future-perfect/2018/11/2/18053418/elon-musk-artificial-intelligence-google-deepmind-openai.
10. "Mohammad Al Gergawi in a Conversation with Elon Musk during WGS17" (February 15, 2017), https://www.youtube.com/watch?v=rCoFKUJ_8Yo&t=1150s.
11. "Elon Musk and Y Combinator President on Thinking for the Future Vanity Fair New Establishment Summit" (October 8, 2015), https://www.youtube.com/watch?v=SqE0107j-uw.
12. "Joe Rogan Experience #1169 – Elon Musk," (September 7, 2018), https://www.youtube.com/watch?v=ycPr5-27vSI&rco=1.
13. Ibid.
14. Ibid.
15. Ibid.
16. Karl Marx, *Capital: A Critique of Political Economy*, vol. 1 (London: Penguin, 1976 [1867]), 742.
17. The shared location was the Pioneer Building at 3180 18th Street in San Francisco's Mission District.
18. "Watch Elon Musk's Original Neuralink Presentation" (July 17, 2019), https://www.youtube.com/watch?v=lA77zsJ3inA.
19. Lex Fridman, "Elon Musk: Neuralink and the Future of Humanity | Lex Fridman Podcast #438" (August 2, 2024), https://www.youtube.com/watch?v=Kbk9BiPhm7o.
20. "Joe Rogan Experience #1169 – Elon Musk" (September 7, 2018), https://www.youtube.com/watch?v=ycPr5-27vSI&rco=1.

21. "Watch Elon Musk's Original Neuralink Presentation" (July 17, 2019), https://www.youtube.com/watch?v=lA77zsJ3inA.
22. Niko McCarty and Milan Cvitkovic, "DARPA Neurotechnology: The Deep Dive," *Asimov Press* (November 8, 2024), https://www.asimov.press/p/darpa-neurotech.
23. Annie Jacobsen, *The Pentagon's Brain: An Uncensored History of DARPA, America's top secret military research agency* (New York: Little, Brown and Company, 2015).
24. Robbin A. Miranda et al., "DARPA-Funded Efforts in the Development of Novel Brain–Computer Interface Technologies," *Journal of Neuroscience Methods*, no. 244 (2015).
25. Geoffrey Ling, "Revolutionizing Prosthetics," (n.d. [2006]), https://web.archive.org/web/20060705131136/https://www.darpa.mil/dso/thrust/biosci/revprost.htm.
26. McCarty and Cvitkovic, "DARPA Neurotechnology: The Deep Dive."
27. Kevin Warwick, *I, Cyborg* (London: Century, 2002); Ray Kurzweil, *The Singularity is Near: When Humans Transcend Biology* (New York: Viking, 2005).
28. "Ray Kurzweil Discusses the Future of BCI (Brain–Computer Interfaces) at a Workshop at the X-Prize Lab in MIT, January 2010," https://www.youtube.com/watch?v=15sho5wrQ6Y.
29. Nick Paton Walsh, "Alter Our DNA or Robots Will Take Over, Warns Hawking," *Guardian* (September 2, 2001), https://www.theguardian.com/uk/2001/sep/02/medicalscience.genetics.
30. https://x.com/ModdedQuad/status/1771230292839145541, March 22, 2024.
31. Jenny Kleeman, "Elon Musk Put a Chip in This Paralysed Man's Brain. Now He Can Move Things With His Mind. Should We Be Amazed – or Terrified?," *Guardian* (February 8, 2025), https://www.theguardian.com/science/2025/feb/08/elon-musk-chip-paralysed-man-noland-arbaugh-chip-brain-neuralink.
32. Robbin A. Miranda et al., "DARPA-funded efforts in the development of novel brain–computer interface technologies."
33. Kleeman, "Elon Musk Put a Chip in This Paralysed Man's Brain."

34. Elon Musk and Neuralink, "An Integrated Brain–Machine Interface Platform with Thousands of Channels," *Journal of Medical Internet Research* 21, no. 10 (2019), https://doi.org/10.2196/16194.
35. AN Pisarchik, VA Maksimenko, and AE Hramov, "From Novel Technology to Novel Applications: Comment on 'An Integrated Brain-Machine Interface Platform With Thousands of Channels' by Elon Musk and Neuralink," *Journal of Medical Internet Research* 21, no. 10 (2019), https://doi.org/10.2196/16356.
36. See Joel E. Dimsdale, *Dark Persuasion: A History of Brainwashing from Pavlov to Social Media* (New Haven: Yale University Press, 2021); Andreas Killen, *Nervous Systems: Brain Science in the Early Cold War* (New York: Harper, 2023).
37. https://x.com/elonmusk/status/1255380013488189440, April 29, 2020.
38. Lauren Feiner, "Elon Musk Says Orders to Stay Home Are 'fascist' in Expletive-Laced Rant During Tesla Earnings Call," *CNBC* (April 29, 2020), https://www.cnbc.com/2020/04/29/elon-musk-slams-coronavirus-shelter-in-place-orders-as-fascist.html.
39. https://x.com/realDonaldTrump/status/1260203080076931072, May 12, 2020. https://x.com/elonmusk/status/1260269880458178560, May 12, 2020.
40. Dana Hull et al., "Musk Reopens Tesla's Plant, Dares Authorities to Arrest Him," Bloomberg (May 11, 2020), https://www.bloomberg.com/news/articles/2020-05-11/tesla-proceeds-with-reopen-newsom-thinks-is-happening-next-week.
41. Faiz Siddiqui, *Hubris Maximus: The Shattering of Elon Musk* (New York: St. Martin's Press, 2025), 79.
42. Higgins, *Power Play*, 330; Lora Kolodny and Michael Wayland, "Tesla Shares Soar After Reporting Big Beat on Second-Quarter Deliveries," *CNBC* (July 2, 2020), https://www.cnbc.com/2020/07/02/tesla-tsla-q2-2020-vehicle-delivery-and-production-numbers.html.
43. Pippa Stevens, "Tesla Tops Toyota to Become Largest Automaker by Market Value," *CNBC* (July 1, 2020), https://www.cnbc.com/2020/07/01/tesla-tops-toyota-to-become-largest-automaker-by-market-value.html.
44. https://x.com/elonmusk/status/1236029449042198528, March 6, 2020.

45. Andrea Fuller et al., "Elon Musk's Hard Turn to Politics, in 300,000 of His Own Words," *Wall Street Journal* (August 25, 2024), https://www.wsj.com/tech/elon-musk-politics-trump-social-media-267d34c8.
46. "Joe Rogan Experience #1609 – Elon Musk" (February 10, 2021), https://www.youtube.com/watch?v=Gbb2rV7Vpnw&t=4624s.
47. https://x.com/elonmusk/status/1243613853558231041, March 27, 2020.
48. Richard Dawkins, "Viruses of the Mind," *Free Inquiry* (Summer 1993): 34-41.
49. "Joe Rogan Experience #1470 – Elon Musk," (May 7, 2020), https://www.youtube.com/watch?v=RcYjXbSJBN8.
50. Larry Buchanan, Quoctrung Bui, and Jugal K. Patel, "Black Lives Matter May Be the Largest Movement in U.S. History," *New York Times* (July 3, 2020), https://www.nytimes.com/interactive/2020/07/03/us/george-floyd-protests-crowd-size.html.
51. Neal Caren, Kenneth T. Andrews, and Micah H. Nelson, "Black Lives Matter (BLM) Protests and the 2020 Presidential Election," *Social Movement Studies* 24, no. 3 (2025); Bouke Klein Teeselink and Georgios Melios, "Weather to Protest: The Effect of Black Lives Matter Protests on the 2020 Presidential Election," *Political Behavior* (2025), https://doi.org/10.1007/s11109-025-10014-w; Diana C. Mutz, "Effects of Changes in Perceived Discrimination During BLM on the 2020 Presidential Election," *Science Advances* 8, no. 9 (2022), https://doi.org/10.1126/sciadv.abj9140.
52. https://x.com/elonmusk/status/1472371245744373760, December 18, 2021.
53. For Rubin's first tweet about the "progressive mind virus" see https://x.com/RubinReport/status/1196545913306656774, November 18, 2019. For his first use of the phrase "Wokeism is a mind virus" see https://x.com/RubinReport/status/1214963616597729280, January 8, 2020. Rubin tweeted the sentence twenty times in the first seven months of 2020.
54. https://x.com/elonmusk/status/1527356085090545664, May 19, 2022.
55. Conger and Mac, *Character Limit*, 99–100; Dana Hull and Jennifer Jacobs, "Tesla, Who? Biden Can't Bring Himself to Say It—and Musk Has Noticed," Bloomberg (February 2, 2022), https://www.

bloomberg.com/news/articles/2022-02-02/tesla-who-biden-can-t-bring-himself-to-say-name-angering-musk; Jesus Mesa, "How Joe Biden Drove Elon Musk Into the Arms of Donald Trump," *Newsweek* (October 13, 2024), https://www.newsweek.com/elon-musk-donald-trump-joe-biden-tesla-spacex-1967931.

56. Isaacson, *Elon Musk*, 420–21; Dana Mattioli and Emily Glazer, "How Elon Musk Broke With Biden and the Democrats," *Wall Street Journal* (July 28, 2024), https://www.wsj.com/politics/elections/how-elon-musk-broke-with-biden-and-the-democrats-4960b7d8.
57. Eric Johnson, Loren Grush, and Malathi Nayak, "SpaceX Discriminated Against Refugees in Hiring, DOJ Says," Bloomberg (August 24, 2023), https://www.bloomberg.com/news/articles/2023-08-24/us-justice-department-sues-spacex-over-discrimination-in-hiring?sref=apOkUyd1. U.S. Department of Justice, "Justice Department Sues SpaceX for Discriminating Against Asylees and Refugees in Hiring," (August 24, 2023), https://www.justice.gov/archives/opa/pr/justice-department-sues-spacex-discriminating-against-asylees-and-refugees-hiring.
58. https://x.com/elonmusk/status/1602278477234728960, December 12, 2022.
59. https://x.com/elonmusk/status/1601277424594927618, December 9, 2022.
60. https://x.com/elonmusk/status/1286869404874088448, July 24, 2020.
61. Ella Yurman, "Vivian Wilson on Being Elon Musk's Estranged Daughter, Going Viral, and Protecting Trans Youth," *Teen Vogue* (March 20, 2025), https://www.teenvogue.com/story/vivian-jenna-wilson-elon-musk-trans-youth; "Elon Musk's Daughter Granted Legal Name, Gender Change," AP News (June 23, 2022), https://apnews.com/article/elon-musk-entertainment-gender-identity-santa-monica-eb64ed75c5e8228ca3cae8e1d836e93d.
62. "Dr. Peterson x Elon Musk," (July 22, 2024), https://x.com/i/broadcasts/1LyGBgPvoDjJN.
63. Donna Haraway, "A Manifesto for Cyborgs: Science, Technology, and Socialist Feminism in the 1980s," *Socialist Review*, no. 80 (1985): 82.

64. Ibid., 93–5.
65. Ibid., 72.
66. Ibid., 66.

7. GODWIN'S ENGINE

1. Board of Governors of the Federal Reserve System (US), "Federal Funds Effective Rate [FEDFUNDS]," https://fred.stlouisfed.org/series/FEDFUNDS; Jeff Cox, "Federal Reserve Approves First Interest Rate Hike in More Than Three Years, Sees Six More Ahead," *CNBC* (March 16, 2022), https://www.cnbc.com/2022/03/16/federal-reserve-meeting.html.
2. Mike Volpi, "How Startups Should Handle the Downturn," *TechCrunch* (June 13, 2022), https://techcrunch.com/2022/06/13/how-startups-should-handle-the-downturn/.
3. Jesse Pound and Samantha Subin, "Stocks Fall to End Wall Street's Worst Year Since 2008, S&P 500 Finishes 2022 Down Nearly 20%," *CNBC* (December 29, 2022), https://www.cnbc.com/2022/12/29/stock-market-futures-open-to-close-news.html.
4. "Layoffs Tracker," https://layoffs.fyi/.
5. Conger and Mac, *Character Limit*, 102.
6. Giles Turner and Maxwell Adler, "Elon Musk Makes $43 Billion Unsolicited Bid to Take Twitter Private," Bloomberg (April 14, 2022), https://www.bloomberg.com/news/articles/2022-04-14/elon-musk-launches-43-billion-hostile-takeover-of-twitter.
7. https://x.com/elonmusk/status/1608919969953349634, December 30, 2022.
8. "FULL Elon Musk Interview with Tucker Carlson FOX News April 2023 (Unedited)" (April 2023), https://www.youtube.com/watch?v=bQ45lsDxL6Q.
9. Ibid.
10. James Clayton, "Elon Full BBC Interview," (April 12, 2023), https://www.youtube.com/watch?v=bRkcLYbvApU.
11. "Joe Rogan Experience #2054 – Elon Musk," (October 30, 2023), https://www.youtube.com/watch?v=tAJUwiAqW38.

12. https://x.com/elonmusk/status/1595250835096621057, November 22, 2022.
13. "Department of Justice Report Regarding the Criminal Investigation into the Shooting Death of Michael Brown by Ferguson, Missouri Police Officer Darren Wilson," (March 4, 2015), https://www.justice.gov/sites/default/files/opa/press-releases/attachments/2015/03/04/doj_report_on_shooting_of_michael_brown_1.pdf.
14. Matt Novak, "Elon Musk Tweets Defense of Cop Who Killed Unarmed Black Man in Ferguson, Missouri," *Gizmodo* (November 23, 2022), https://gizmodo.com/elon-musk-tweets-cop-killed-unarmed-black-man-ferguson-1849815713.
15. https://x.com/elonmusk/status/1595535360863395842, November 23, 2022.
16. Curtis Yarvin, "The Twitter Coup," *Gray Mirror* (April 15, 2022), https://graymirror.substack.com/p/the-twitter-coup.
17. Brandy Zadrozny, "Elon Musk's 'Amnesty' Pledge Brings Back QAnon, Far-Right Twitter Accounts," *NBC News* (December 2, 2022), https://www.nbcnews.com/tech/internet/elon-musks-twitter-beginning-take-shape-rcna58940.
18. Paolo Gerbaudo, *The Digital Party: Political Organisation and Online Democracy* (London: Pluto Press, 2019).
19. Zoë Schiffer and Casey Newton, "Yes, Elon Musk Created a Special System for Showing You All His Tweets First," *The Verge* (February 15, 2023), https://www.theverge.com/2023/2/14/23600358/elon-musk-tweets-algorithm-changes-twitter; Timothy Graham and Mark Andrejevic, "Tech Billionaire Elon Musk's Social Media Posts Have Had a 'Sudden Boost' since July, New Research Reveals," *The Conversation* (October 31, 2024), https://theconversation.com/tech-billionaire-elon-musks-social-media-posts-have-had-a-sudden-boost-since-july-new-research-reveals-242490.
20. "About X Premium," https://help.x.com/en/using-x/x-premium.
21. David Ingram and Bruna Horvath, "How Elon Musk is Boosting Far-Right Politics Across the Globe," *NBC News* (February 16, 2025), https://www.nbcnews.com/tech/elon-musk/elon-musk-boosting-far-right-politics-globe-rcna189505.

22. https://x.com/geertwilderspvv/status/1742596599631446238, January 3, 2024; https://x.com/elonmusk/status/1742826319849684997, January 4, 2024.
23. https://x.com/elonmusk/status/1945546893385175292, July 16, 2025.
24. For example: https://x.com/elonmusk/status/1961844695547494661, August 30, 2025. https://x.com/elonmusk/status/1962903682997182528, September 2, 2025. https://x.com/elonmusk/status/1875620755104526755, January 4, 2025. https://x.com/elonmusk/status/1875145167633887358, January 3, 2025.
25. Musk posted the tweet on March 8, 2024 but subsequently deleted it. See Sohrab Ahmari, "The New Racist Right are Uniquely Dangerous," *New Statesman*, April 10, 2024, https://www.newstatesman.com/comment/2024/04/the-new-racist-right-are-uniquely-dangerous.
26. https://x.com/elonmusk/status/1625580694494928897, February 14, 2023.
27. José Pedro Zúquete, *The Identitarians: The Movement against Globalism and Islam in Europe* (Notre Dame, Indiana: University of Notre Dame Press,, 2018), 146–59.
28. https://x.com/elonmusk/status/1724908287471272299, November 15, 2023; Mike Wendling, "White House Criticises Elon Musk over 'Hideous' Antisemitic Lie," *BBC* (November 17, 2023), https://www.bbc.com/news/world-us-canada-67446800.
29. Nick Seaver, *Computing Taste: Algorithms and The Makers of Music Recommendation* (Chicago: University of Chicago Press, 2022).
30. Erika Kinetz and Aaron Kessler, "Musk, a Social Media Powerhouse, Boosts Fortunes of Hard-Right Figures in Europe," AP News (August 1, 2025), https://apnews.com/article/musk-europe-politicians-influence-x-twitter-extremists-89746e1e17bcc134206c14a204efcbce.
31. Calder McHugh, "The German Musk Whisperer," *Politico* (May 8, 2025), https://www.politico.com/newsletters/politico-nightly/2025/05/08/the-german-musk-whisperer-00336986.
32. "Elon Musk–Alice Weidel Full Conversation: Tesla CEO Speaks to German Far-Right Party AfD Chief," *The Economic Times*

(January 9, 2025), https://www.youtube.com/watch?v=cpjKbWKZnoo.
33. Rachel Treisman, "Elon Musk Faces Criticism for Encouraging Germans to Move Beyond 'past guilt'," *NPR* (January 27, 2025), https://www.npr.org/2025/01/27/nx-s1-5276084/elon-musk-german-far-right-afd-holocaust.
34. https://x.com/elonmusk/status/1962406618886492245, September 1, 2025.
35. Ingram and Horvath, "How Elon Musk is Boosting Far-Right Politics Across the Globe."
36. Hao, *Empire of AI*, 260.
37. Stephen Morris and Rafe Uddin, "Big Tech Lines Up Over $300bn in AI Spending for 2025," *Financial Times* (February 7, 2025), https://www.ft.com/content/634b7ec5-10c3-44d3-ae49-2a5b9ad566fa.
38. Mike Isaac, "Silicon Valley Is in Its 'Hard Tech' Era," *New York Times* (August 4, 2025), https://www.nytimes.com/2025/08/04/technology/ai-silicon-valley-hard-tech.html.
39. Hao, *Empire of AI*, 62–5.
40. https://x.com/elonmusk/status/1603836383885332480, December 16, 2022.
41. Dr. Peterson x Elon Musk" (July 22, 2024), https://x.com/i/broadcasts/1LyGBgPvoDjJN.
42. "Joe Rogan Experience #2281 – Elon Musk."
43. Ibid.
44. "FULL Elon Musk Interview with Tucker Carlson FOX News April 2023 (Unedited)," (April 2023), https://www.youtube.com/watch?v=bQ45lsDxL6Q.
45. https://x.com/xai/status/1721027348970238035, November 5, 2023.
46. https://x.com/elonmusk/status/1866384370044686522, December 10, 2024. https://x.com/elonmusk/status/1866384589784006738, December 10, 2024. https://x.com/elonmusk/status/1866485009789681907, December 10, 2024.
47. "Pepe the Frog Meme Branded a 'hate symbol'."
48. Musk's initial offer for Twitter in April 2022 valued the company at $43 billion, but the final deal that closed in October 2022 was worth $44 billion. Greg Bensinger, "Musk's Social Media Firm

X Bought by His AI Company, Valued at $33 billion," *Reuters* (March 29, 2025), https://www.reuters.com/markets/deals/musks-xai-buys-social-media-platform-x-45-billion-2025-03-28/.
49. "Colossus," https://x.ai/colossus.
50. Southern Environmental Law Center, "Elon Musk's xAI Facility is Using Gas Turbines in South Memphis, We're Taking Action," (September 17, 2025), https://www.selc.org/news/resistance-against-elon-musks-xai-facility-in-south-memphis-gets-stronger/; Wendi C. Thomas, "Inside the Memphis Chamber of Commerce's Push for Elon Musk's xAI Data Center," ProPublica (August 22, 2025), https://www.propublica.org/article/memphis-xai-colossus-elon-musk-chamber-messaging.
51. KeShaun Pearson post on Facebook, (April 25, 2025), https://www.facebook.com/keshaun.pearson/posts/a-decade-of-continued-failure-has-to-end-here-with-this-illegal-pollution-of-our/10228081910653594/.
52. Isaacson, *Elon Musk*, 602.
53. Yanis Varoufakis, "The Techno-Feudal Method to Musk's Twitter Madness," *Project Syndicate* (November 24, 2022), https://www.project-syndicate.org/commentary/musk-bought-twitter-to-get-cloud-capital-by-yanis-varoufakis-2022-11.
54. https://x.com/elonmusk/status/1658334514462982144, May 16, 2023.
55. Grace Kay, "Inside Grok's War On 'Woke'," *Business Insider* (February 28, 2025), https://www.businessinsider.com/xai-grok-training-bias-woke-idealogy-2025-02.
56. Ibid.
57. Ibid.
58. https://x.com/elonmusk/status/1936493967320953090, June 21, 2025; Kelsey Piper, "Grok's MechaHitler disaster is a preview of AI disasters to come," *Vox* (July 11, 2025), https://www.vox.com/future-perfect/419631/grok-hitler-mechahitler-musk-ai-nazi.
59. Kate Conger, "Employee's Change Caused xAI's Chatbot to Veer Into South African Politics," *New York Times* (May 16, 2025), https://www.nytimes.com/2025/05/16/technology/xai-elon-musk-south-africa.html.

60. Kate Conger, "Grok Chatbot Mirrored X Users' 'Extremist Views' in Antisemitic Posts, xAI Says," *New York Times* (July 12, 2025), https://www.nytimes.com/2025/07/12/technology/x-ai-grok-antisemitism.html.
61. Amy Kraft, "Microsoft Shuts Down AI Chatbot After it Turned into a Nazi," *CBS News* (March 25, 2016), https://www.cbsnews.com/news/microsoft-shuts-down-ai-chatbot-after-it-turned-into-racist-nazi.
62. https://x.com/elonmusk/status/721946756984938497, April 18, 2016.
63. Mike Godwin, "Meme, Counter-meme," *Wired* (October 1, 1994), https://www.wired.com/1994/10/godwin-if-2/.
64. When tested, it turned out not to be true. Gabriele Fariello, Dariusz Jemielniak, and Adam Sulkowski, "Does Godwin's Law (Rule of Nazi Analogies) Apply in Observable Reality?: An Empirical Study of Selected Words in 199 Million Reddit Posts," *New Media & Society* 26, no. 1 (2024): 389-404.
65. Stuart A. Thompson et al., "How Elon Musk Is Remaking Grok in His Image," *New York Times* (September 2, 2025), https://www.nytimes.com/2025/09/02/technology/elon-musk-grok-conservative-chatbot.html.
66. https://x.com/elonmusk/status/1935180620352958935, June 17, 2025.
67. https://x.com/elonmusk/status/1944132781745090819, July 12, 2025.
68. Isaacson, *Elon Musk*, 418.
69. Matteo Wong, "What Elon Musk's Version of Wikipedia Thinks About Hitler, Putin, and Apartheid," *The Atlantic* (October 28, 2025), https://www.theatlantic.com/technology/2025/10/grokipedia-elon-musk/684730/; Hadas Gold, "Elon Musk Launches His Version of Wikipedia," *CNN* (October 28, 2025), https://www.cnn.com/2025/10/28/tech/elon-musk-launches-grokipedia-wikipedia.
70. https://x.com/AutismCapital/status/1942982539997798415, July 9, 2025.
71. Haraway, "A Manifesto for Cyborgs: Science, Technology, and Socialist Feminism in the 1980s," 79.

8. STATE X

1. "TRANSCRIPT: Elon Musk on Joe Rogan Experience Podcast #2281" (February 28, 2025), https://singjupost.com/transcript-elon-musk-on-joe-rogan-experience-podcast-2281/.
2. "Transcript of Elon Musk on Verdict with Senator Ted Cruz Podcast (Part 1)," *The Singju Post* (March 20, 2025), https://singjupost.com/transcript-of-elon-musk-on-verdict-with-senator-ted-cruz-podcast-part-1/.
3. "Establishing and Implementing the President's 'Department of Government Efficiency'," *The White House Executive Order* (January 20, 2025), https://www.whitehouse.gov/presidential-actions/2025/01/establishing-and-implementing-the-presidents-department-of-government-efficiency/.
4. Tim O'Reilly, "Government as a Platform," *innovations* 6, no. 1 (2010): 37.
5. Stephanie Ricker Schulte, "United States Digital Service: How 'Obama's Startup' Harnesses Disruption and Productive Failure to Reboot Government," *International Journal of Communication* 12 (2018): 3.
6. Helen Margetts and Patrick Dunleavy, "The Political Economy of Digital Government: How Silicon Valley Firms Drove Conversion to Data Science and Artificial Intelligence in Public Management," *Public Money & Management* (2024), https://doi.org/10.1080/09540962.2024.2389915.
7. See Brendan McQuade, *Pacifying the Homeland: Intelligence Fusion and Mass Supervision* (Berkeley, CA: University of California Press, 2019).
8. See Henry Farrell and Abraham Newman, *Underground Empire: How America Weaponized the World Economy* (New York: Holt, 2023), chapter 2.
9. Glenn Greenwald, "XKeyscore: NSA Tool Collects 'nearly everything a user does on the internet'," *Guardian* (July 31, 2013), https://www.theguardian.com/world/2013/jul/31/nsa-top-secret-program-online-data.
10. Andrej Zwitter, "Cybernetic Governance: Implications of Technology Convergence on Governance Convergence," *Ethics and Information Technology* 26, no. 24 (2024): 3, https://doi.org/10.1007/s10676-024-09763-9.

11. Sophie Alexander and Jamie Tarabay, "Peter Thiel's Deep Ties to Trump's Top Ranks," Bloomberg (March 7, 2025), https://www.bloomberg.com/features/2025-peter-thiel-trump-administration-connections/.
12. https://x.com/elonmusk/status/1825713824785379477, August 19, 2024; "How Musk Built DOGE: Timeline and Key Takeaways," *New York Times* (February 28, 2025), https://www.nytimes.com/2025/02/28/us/politics/musk-doge-timeline-takeaways.html.
13. https://x.com/elonmusk/status/1887177695789760647, February 5, 2025; Joe Wilkins, "DOGE.gov Website Launches With Mangled, AI-Generated American Flag," *Yahoo News* (January 22, 2025), https://www.yahoo.com/news/doge-gov-website-launches-mangled-155304011.html.
14. "CPAC Interview with DOGE Chair Elon Musk," *C-Span* (February 20, 2025), https://www.c-span.org/program/public-affairs-event/cpac-interview-with-doge-chair-elon-musk/656121.
15. "Joe Rogan Experience #2281 – Elon Musk."
16. Kirsten Grind and Megan Twohey, "On the Campaign Trail, Elon Musk Juggled Drugs and Family Drama," *New York Times* (May 30, 2025), https://www.nytimes.com/2025/05/30/us/elon-musk-drugs-children-trump.html.
17. https://x.com/elonmusk/status/1713702870699086052, October 15, 2023.
18. https://x.com/elonmusk/status/1902642215928656206, March 20, 2025.
19. Ethan Gach, "Elon Musk Installed A Gaming PC At His DOGE Office," *Kotaku* (February 28, 2025), https://kotaku.com/elon-musk-gaming-pc-doge-office-diablo-path-of-exile-1851767110.
20. The statement was posted as a reply to a video of a worm-like parasite leaving the body of a praying mantis as it drowned. https://x.com/elonmusk/status/1915154454250479703, April 23, 2025.
21. https://x.com/elonmusk/status/1861475058428096875, November 26, 2024.
22. https://x.com/elonmusk/status/1306491844512026624, September 17, 2020.

23. Scott Patterson, Josh Dawsey, and Brian Schwartz, "Inside DOGE's Clash With the Federal Workforce," *Wall Street Journal* (February 27, 2025), https://www.wsj.com/politics/policy/inside-doge-elon-musk-government-employees-b87fc17a.
24. Doina Chiacu, "Musk Wants to 'delete entire agencies' From US Government," Reuters (February 13, 2025), https://www.reuters.com/world/us/musk-wants-delete-entire-agencies-us-government-2025-02-13/.
25. Benjamin Wallace-Wells, "What did Elon Musk Accomplish at DOGE?," *The New Yorker* (June 16, 2025), https://www.newyorker.com/magazine/2025/06/23/what-did-elon-musk-accomplish-at-doge.
26. Emily Tavoulareas, "DOGE Understands Something the US Policy Establishment Does Not: Technology is the Spinal Cord of Government," *Tech Policy Press* (February 18, 2025).
27. Makena Kelly et al., "Inside Elon Musk's 'Digital Coup'," *Wired* (March 13, 2025), https://www.wired.com/story/elon-musk-digital-coup-doge-data-ai/.
28. Conger and Mac, *Character Limit*, 370; Arnaud Leparmentier, "Musk's Zero-Based Budgeting: Cut Everything, Rebuild What's Essential," *Le Monde* (February 18, 2025), https://www.lemonde.fr/en/opinion/article/2025/02/18/musk-s-zero-based-budgeting-cut-everything-rebuild-what-s-essential_6738289_23.html; Hannah Natanson et al., "Move Fast, Break Things, Rebuild: Elon Musk's Strategy for U.S. Government," *Washington Post* (February 24, 2025), https://www.washingtonpost.com/technology/2025/02/24/doge-fast-cuts-federal-workers-programs-elon-musk.
29. Seth Catalli, "The Problem that Made 'Zero-Based Budgeting' Unachievable Just Got Solved," *CFO Dive* (May 14, 2024), https://www.cfodive.com/news/the-problem-that-made-zero-based-budgeting-unachievable-just-got-solved/715611/.
30. Kelly et al., "Inside Elon Musk's 'Digital Coup'."
31. https://x.com/elonmusk/status/1886307316804263979, February 3, 2025.
32. Rodney Coyte, Martin Messner, and Shan Zhou, "The Revival of Zero-Based Budgeting: Drivers and Consequences of Firm-Level Adoptions," *Accounting & Finance* 62, no. 3 (Sep 2022): 3182.

33. Eryk Salvaggio, "Anatomy of an AI Coup," *Tech Policy Press* (February 9, 2025), https://www.techpolicy.press/anatomy-of-an-ai-coup/.
34. "Stopping Waste, Fraud, and Abuse by Eliminating Information Silos," *The White House Executive Order* (March 20, 2025), https://www.whitehouse.gov/presidential-actions/2025/03/stopping-waste-fraud-and-abuse-by-eliminating-information-silos/.
35. Makena Kelly, "DOGE Is Planning a Hackathon at the IRS. It Wants Easier Access to Taxpayer Data," *Wired* (April 5, 2025), https://www.wired.com/story/doge-hackathon-irs-data-palantir/.
36. Conger and Mac, *Character Limit*, 195.
37. See Charlie Warzel, Ian Bogost, and Matteo Wong, "DOGE has 'God Mode' Access to Government Data," *The Atlantic* (February 19, 2025).
38. Makena Kelly, "Palantir Is Extending Its Reach Even Further Into Government," *Wired* (August 1, 2025), https://www.wired.com/story/palantir-government-contracting-push/; Sheera Frenkel and Aaron Krolik, "Trump Taps Palantir to Compile Data on Americans," *New York Times* (May 30, 2025), https://www.nytimes.com/2025/05/30/technology/trump-palantir-data-americans.html.
39. See Natanson et al., "Move Fast, Break Things, Rebuild."
40. Vittoria Elliott, Leah Feiger, and Tim Marchman, "The US Treasury Claimed DOGE Technologist Didn't Have 'Write Access' When He Actually Did," *Wired* (February 6, 2025), https://www.wired.com/story/treasury-department-doge-marko-elez-access/; Brandon Roberts and Vernal Coleman, "Inside the AI Prompts DOGE Used to 'Munch' Contracts Related to Veterans' Health," ProPublica (June 6, 2025), https://www.propublica.org/article/inside-ai-tool-doge-veterans-affairs-contracts-sahil-lavingia.
41. Jared Keller, "The US Army Is Using 'CamoGPT' to Purge DEI From Training Materials," *Wired* (March 8, 2025), https://www.wired.com/story/the-us-army-is-using-camogpt-to-purge-dei-from-training-materials/; Ariana Baio, "Doge Used Flawed AI Tool to 'Munch' Veteran Affairs Contracts, Report Claims,"

Independent (June 6, 2025), https://www.independent.co.uk/news/world/americas/us-politics/doge-flawed-ai-veteran-affairs-b2765364.html.

42. Hannah Natanson et al., "DOGE Builds AI Tool to Cut 50 Percent of Federal Regulations," *Washington Post* (July 26, 2025), https://www.washingtonpost.com/business/2025/07/26/doge-ai-tool-cut-regulations-trump/.
43. Zeynep Tufekci, "Here Are the Digital Clues to What Musk Is Really Up To," *New York Times* (February 21, 2025), https://www.nytimes.com/2025/02/21/opinion/musk-doge-personal-data.html.
44. Todd C. Frankel and Hannah Natanson, "Why DOGE is Struggling to Find Fraud in Social Security," *Washington Post* (March 24, 2025), https://www.washingtonpost.com/business/2025/03/24/social-security-fraud-doge-cuts-dead/.
45. https://x.com/elonmusk/status/1891350795452654076, February 16, 2025.
46. Mishal Husain, "Elon Musk on Political Spending: 'I Think I've Done Enough'," Bloomberg (May 21, 2025), https://www.bloomberg.com/features/2025-elon-musk-weekend-interview/. See also Charlie Warzel and Hana Kiros, "Elon Musk is Playing God," *The Atlantic* (June 24, 2025), https://www.theatlantic.com/technology/archive/2025/06/elon-musk-usaid-cuts/683299/.
47. "Joe Rogan Experience #2281 – Elon Musk."
48. Walter Isaacson, *Elon Musk* (New York: Simon & Schuster, 2023), 426.
49. Matthew Purdy, "The Techno-Futuristic Philosophy Behind Elon Musk's Mania," *New York Times* (May 29, 2025), https://www.nytimes.com/2025/05/29/business/elon-musk-longtermism-effective-altruism-doge.html.
50. Nick Bostrom, "Are We Living in a Computer Simulation?," *The Philosophical Quarterly* 53, no. 211 (Apr 2003): 254.
51. Robin Hanson, "How to Live in a Simulation," *Journal of Evolution and Technology* 7 (2001), https://mason.gmu.edu/~rhanson/Lifeinsim.html.
52. Ben Werschkul, "How Elon Musk's Time at PayPal Shaped His Approach to Overhauling Social Security," *Yahoo! Finance* (March 29, 2025), https://ca.finance.yahoo.com/news/

how-elon-musks-time-at-paypal-shaped-his-approach-to-overhauling-social-security-150048508.html; Conger and Mac, *Character Limit*, 286.

53. https://x.com/elonmusk/status/1455233931880587267, November 1, 2021.
54. https://x.com/elonmusk/status/1633966756107624448, March 9, 2023.
55. https://x.com/elonmusk/status/1637864049307430914, March 20, 2023.
56. https://x.com/elonmusk/status/1846783171012002005, October 17, 2024.
57. https://x.com/elonmusk/status/1902602137332257043, March 20, 2025.
58. https://x.com/elonmusk/status/1644036581466992658, April 6, 2023.
59. https://x.com/elonmusk/status/1766544406532780200, March 9, 2024.
60. https://x.com/elonmusk/status/1876446327254704534, January 6, 2025; https://x.com/elonmusk/status/1889194985066475520, February 11, 2025.
61. https://x.com/elonmusk/status/1753590787130994745, February 2, 2024.
62. https://x.com/elonmusk/status/1707146779894951982, September 27, 2023.
63. https://x.com/elonmusk/status/1708197042529697947, September 30, 2023.
64. https://x.com/elonmusk/status/1708585414708109576, October 1, 2023.
65. https://x.com/elonmusk/status/1768078994161672303, Mar 13, 2024; https://x.com/elonmusk/status/1788431023593697519, May 9, 2024.
66. "Full Transcript—Joe Rogan Experience #2223 – Elon Musk," (November 4, 2024).
67. https://x.com/elonmusk/status/1752185691168182669, January 29, 2024.
68. Alba et al., "Elon Musk Is Now X's Biggest Promoter of Anti-Immigrant Conspiracies."
69. "Elon Musk's Transformation, in His Own Words," *The Economist* (November 21, 2024), https://www.economist.com/briefing/2024/11/21/elon-musks-transformation-in-his-own-words.
70. Elliott, Feiger, and Marchman, "The US Treasury Claimed DOGE Technologist Didn't Have 'Write Access' When He Actually Did";

Katherine Long, "DOGE Staffer Resigns Over Racist Posts," *Wall Street Journal* (February 7, 2025), https://www.wsj.com/tech/doge-staffer-resigns-over-racist-posts-d9f11a93.

71. Makena Kelly and Vittoria Elliott, "DOGE Is Building a Master Database to Surveil and Track Immigrants," *Wired* (April 18, 2025), https://www.wired.com/story/doge-collecting-immigrant-data-surveil-track/.

72. Caroline Haskins, "ICE Is Paying Palantir $30 Million to Build 'ImmigrationOS' Surveillance Platform," *Wired* (April 18, 2025), https://www.wired.com/story/ice-palantir-immigrationos.

73. Alexandra Berzon et al., "Social Security Lists Thousands of Migrants as Dead to Prompt Them to 'Self-Deport,'" *New York Times* (April 10, 2025), https://www.nytimes.com/2025/04/10/us/politics/migrants-deport-social-security-doge.html.

74. Tyler Pager, "Trump Demands Census Excluding Undocumented Immigrants Amid Redistricting Fight," *New York Times* (August 7, 2025), https://www.nytimes.com/2025/08/07/us/politics/trump-census-undocumented-immigrants.html.

75. https://x.com/elonmusk/status/1865968958693618016, December 8, 2024; "Transcript of Elon Musk on Verdict with Senator Ted Cruz Podcast (Part 1)," *The Singju Post* (March 20, 2025), https://singjupost.com/transcript-of-elon-musk-on-verdict-with-senator-ted-cruz-podcast-part-1/.

76. https://x.com/elonmusk/status/1262076474565242880, May 17, 2020.

77. Donna Zuckerberg, *Not All Dead White Men: Classics and Misogyny in the Digital Age* (Cambridge, MA: Harvard University Press, 2018), 14-15.

78. Robert Reich, "The Trump-Vance-Musk-Putin Manosphere," (March 3, 2025), https://robertreich.substack.com/p/the-trump-vance-musk-putin-manosphere.

79. Jamie White, "EPIC Interview: Andrew Tate & Alex Jones Join Forces To Wargame The New World Order's Demise & Break the Matrix!," *Infowars* (August 25, 2023), https://www.infowars.com/posts/epic-interview-andrew-tate-alex-jones-join-forces-to-wargame-the-new-world-orders-demise-break-the-matrix/.

80. The jokey caption read "This is what happens if you take DayQuil & NyQuil at the same time." https://x.com/elonmusk/status/1907313533995368878, April 2, 2025.
81. Joe Miller and Chris Cook, "What has Elon Musk's Doge Actually Achieved?" *Financial Times* (May 14, 2025), https://www.ft.com/content/085430ab-27fe-46fc-a798-1059649d3b32.
82. Jessica Riedl, "The Actual Math Behind DOGE's Cuts," *The Atlantic* (May 8, 2025), https://www.theatlantic.com/politics/archive/2025/05/musk-doge-spending-cuts/682736/.
83. "CBS News Sunday Morning with Jane Pauley," (June 1, 2025), https://www.podchaser.com/podcasts/cbs-news-sunday-morning-with-j-13199/episodes/elon-musk-bill-clinton-the-pen-253693213/transcript.
84. Lydia Saad, "Pope Leo Most Favorably Viewed of 14 Newsmakers," *Gallup* (August 5, 2025), https://news.gallup.com/poll/693155/pope-leo-favorably-viewed-newsmakers.aspx. See poll tracker at https://today.yougov.com/topics/economy/trackers/fame-and-popularity-elon-musk.
85. https://x.com/xai/status/1944776899420377134, July 14, 2025.
86. https://x.com/elonmusk/status/1944705383874146513, July 14, 2025; David Ingram, "Musk's Grok 'Companions' Include a Flirty Anime Character and an Anti-Religion Panda," *NBC News* (July 15, 2025), https://www.nbcnews.com/tech/internet/grok-companions-include-flirty-anime-waifu-anti-religion-panda-rcna218797. Miles Klee, "Grok Rolls Out Pornographic Anime Companion, Lands Department of Defense Contract," *Rolling Stone* (July 14, 2025), https://www.rollingstone.com/culture/culture-news/grok-pornographic-anime-companion-department-of-defense-1235385034/.

CONCLUSION. FOUR FUTURES FOR MUSKISM

1. Karl Polanyi, *The Great Transformation* (New York: Farrar & Rinehart inc., 1944), 226.
2. Quoted in Gareth Dale, *Karl Polanyi: A Life on the Left* (New York: Columbia University Press, 2016), 59.

3. David A. Fahrenthold and Ryan Mac, "Elon Musk Has a Giant Charity. Its Money Stays Close to Home," *New York Times* (March 10, 2024), https://www.nytimes.com/2024/03/10/us/elon-musk-charity.html.
4. Tesla, "Master Plan Part IV," (2025), https://www.tesla.com/master-plan-part-4.
5. Katherine Hamilton, "Elon Musk Says Optimus Robots to Make Up About 80% of Tesla's Value," *Wall Street Journal* (September 2, 2025), https://www.wsj.com/tech/elon-musk-says-optimus-robots-to-make-up-about-80-of-teslas-value-4584a616?st=NxJX41.
6. "Tesla, Inc. (TSLA) Q3 2025 Earnings Call Transcript," (October 22, 2025), https://seekingalpha.com/article/4832142-tesla-inc-tsla-q3-2025-earnings-call-transcript.
7. "Elon Musk: Digital Superintelligence, Multiplanetary Life, How to Be Useful (Transcript)," (June 20, 2025), https://singjupost.com/transcript-of-elon-musk-digital-superintelligence-multiplanetary-life-how-to-be-useful/.
8. Kate Crawford, "Eating the Future: The Metabolic Logic of AI Slop," *e-flux* (September 2025), https://www.e-flux.com/architecture/intensification/6782975/eating-the-future-the-metabolic-logic-of-ai-slop.
9. Alexander C. Karp and Nicholas W. Zamiska, *The Technological Republic: Hard Power, Soft Belief, and the Future of the West* (New York: Crown Currency, 2025).
10. Micah Maidenberg and Becky Peterson, "SpaceX Pushes to Get Starship Rocket Ready for Mars by Next Year," *Wall Street Journal* (May 26, 2025), https://www.wsj.com/science/space-astronomy/spacex-starship-mars-military-elon-musk-3240c18d; Drew FitzGerald and Micah Maidenberg, "Elon Musk's SpaceX Set to Win $2 Billion Pentagon Satellite Deal," *Wall Street Journal* (October 31, 2025), https://www.wsj.com/politics/national-security/elon-musks-spacex-set-to-win-2-billion-pentagon-satellite-deal-c0a51325.
11. https://x.com/elonmusk/status/1931806069220954360, June 8, 2025.
12. Hilaire Belloc, *The Modern Traveller* (New York: Alfred A. Knopf, 1928), 41.

13. Daniel R. Headrick, "The Tools of Imperialism: Technology and the Expansion of European Colonial Empires in the Nineteenth Century," *Journal of Modern History* 51, no. 2 (Jun 1979): 259.
14. Aneesa Ahmed, "Elon Musk Calls for Dissolution of Parliament at Far-Right Rally in London," *Guardian* (September 13, 2025), https://www.theguardian.com/technology/2025/sep/13/elon-musk-calls-for-dissolution-of-parliament-at-far-right-rally-in-london; "Elon Musk's Speech: 'Violence Is Coming' At London Anti-Immigration Rally," *Singju Post* (September 15, 2025), https://singjupost.com/elon-musks-speech-violence-is-coming-at-london-anti-immigration-rally/.
15. "Elon Musk on Why He Wants More Robots and Less Government," *Wall Street Journal* (December 7, 2021), https://www.wsj.com/podcasts/the-journal/elon-musk-on-why-he-wants-more-robots-and-less-government/274b7aec-c858-4be8-b719-6a34bdea86f6.
16. https://x.com/elonmusk/status/1863995164475129976, December 3, 2024.
17. "Elon Musk interviewed at Atreju 2023 (Rome)," (December 16, 2023), https://www.youtube.com/watch?v=WPf4_5DCP-E.
18. https://x.com/elonmusk/status/1918840576834781667, May 3, 2025.
19. https://x.com/elonmusk/status/1964582769302045121, September 7, 2025.
20. https://x.com/elonmusk/status/1962680097816879208, September 1, 2025.
21. Jennifer Dabbs Sciubba, *8 Billion and Counting: How Sex, Death, and Migration Shape Our World* (New York, NY: W.W. Norton & Company, 2022), 2. On Musk's population concerns see Sophie Alexander and Dana Hull, "Elon Wants You to Have More Babies," Bloomberg Businessweek (June 21, 2024), https://www.bloomberg.com/features/2024-elon-musk-population-collapse-baby-push; Quinn Slobodian, "Elon Musk Wants Us to Have More Children," *New Statesman* (July 29, 2024), https://www.newstatesman.com/ideas/2024/07/elon-musk-wants-us-have-more-children-demography-will-macaskill.

22. Susanne Maria Klausen, *Abortion under Apartheid: Nationalism, Sexuality, and Women's Reproductive Rights in South Africa* (New York, NY: Oxford University Press, 2015), 183.
23. Zolan Kanno-Youngs et al., "The Road to Trump's Embrace of White South Africans," *New York Times* (May 14, 2025), https://www.nytimes.com/2025/05/14/us/politics/trump-south-africa-afrikaners.html.
24. Klausen, *Abortion under Apartheid*, 181–3.
25. Mark Harris, "First Space, Then Auto—Now Elon Musk Quietly Tinkers with Education," *Ars Technica* (June 25, 2018), https://arstechnica.com/science/2018/06/first-space-then-auto-now-elon-musk-quietly-tinkers-with-education/.
26. Lauren McGaughy, "An Education Ecosystem is Being Built in Elon Musk's Image. It Starts in Rural Texas," *KUT News* (January 13, 2025), https://www.kut.org/education/2025-01-13/elon-musk-ad-astra-school-education-bastrop-austin-texas. See also Brian Highsmith, "Governing the Company Town" *Stanford Law Review* 77 (June 2025).
27. Greg Grandin, *Fordlandia: The Rise and Fall of Henry Ford's Forgotten Jungle City* (New York: Metropolitan Books, 2009).
28. Sophie Alexander and Dana Hull, "Elon Musk Is Planning a New University in Austin," Bloomberg (December 13, 2023), https://www.bloomberg.com/news/articles/2023-12-13/musk-planning-new-university-in-austin-with-100-million-gift.
29. https://x.com/GregAbbott_TX/status/1735709632218186235, December 15, 2023.
30. https://x.com/elonmusk/status/1754067365707563045, February 4, 2024. It was a recycled joke. Ananya Bhattacharya, "In One Tweet, Elon Musk Captures the Everyday Sexism Faced by Women in STEM," *Quartz* (November 1, 2021), https://qz.com/work/2082746/elon-musks-tweet-captures-everyday-sexism-faced-by-women-in-stem.
31. Grind and Twohey, "On the Campaign Trail, Elon Musk Juggled Drugs and Family Drama"; Juliana Kaplan and Jack Newsham, "A New DOGE Staffer at the Department of Labor has Helped Run a Fertility Clinic and has Pronatalist Ties," *Business*

Insider (March 5, 2025), https://www.businessinsider.com/doge-staffer-fertility-clinic-pronatalist-department-of-labor.

32. Malcolm Collins, Hath, and Simone H. Collins, "New Cause Area: Demographic Collapse," *Effective Altruism Forum* (June 30, 2022), https://forum.effectivealtruism.org/posts/vFfoqL74kmZbydKjp/new-cause-area-demographic-collapse.

33. Quoted in Dana Mattioli, "The Tactics Elon Musk Uses to Manage His 'Legion' of Babies—and Their Mothers," *Wall Street Journal* (April 15, 2025), https://www.wsj.com/politics/elon-musk-children-mothers-ashley-st-clair-grimes-dc7ba05c?st=vJ2t5b. Musk tweeted in response: "TMZ >> WSJ." https://x.com/elonmusk/status/1912355094173217172?s=20, April 15, 2025.

34. "CPAC Interview with DOGE Chair Elon Musk."

35. The event was the Y Combinator AI Startup School. "Elon Musk: Digital Superintelligence, Multiplanetary Life, How to Be Useful (Transcript)."

36. "Elon Musk: Digital Superintelligence, Multiplanetary Life, How to Be Useful (Transcript)."

37. "Tesla, Inc. (TSLA) Q3 2025 Earnings Call Transcript."

38. Musk has pledged $1m for murals of murder victim Iryna Zarutska. Isabel Keane, "Musk Pledges $1 Million for Memorials Honoring Ukrainian Refugee Murdered on US Train," *Independent* (September 11, 2025), https://www.independent.co.uk/news/world/americas/elon-musk-donation-ukraine-refugee-murder-b2824659.html.

39. Musk posted: "History will call this 'The Rape of Europe.'" https://x.com/elonmusk/status/1961844695547494661, August 30, 2025.

40. "Is it fair to say 'the U.K. is being invaded'...?" Musk responded, "Obviously." https://x.com/elonmusk/status/1962313892790436130, August 31, 2025. In response to a tweet sharing a headline from the *Daily Mail* about an "Iranian man who raped his lodger in London," Musk responded, "Suicidal empathy." https://x.com/elonmusk/status/1962215679869833522, August 31, 2025. The next day, Musk posted: "Remigration is the only way." https://x.com/elonmusk/status/1962406618886492245, September 1, 2025. The day after that, Musk retweeted a tweet from Eva Vlaardingerbroek

saying "WE ARE GENERATION REMIGRATION." https://x.com/elonmusk/status/1962984560456864226, September 2, 2025.
41. Musk posted: "The word 'slave' literally originates from white person, because so many white people were enslaved." https://x.com/elonmusk/status/1962614230223593954, September 1, 2025.
42. https://x.com/elonmusk/status/1926427531726500288, May 24, 2025.
43. Musk posted: "The axiomatic error undermining much of Western Civilization is 'weak makes right.'" https://x.com/elonmusk/status/1783727565989134488, April 26, 2024.
44. Musk posted: "In order to survive, an ideology that fails to reproduce itself must necessarily infect the minds of the children of those who do reproduce. That's why they push so hard to control education. It is a vampire cult." https://x.com/elonmusk/status/1961452078423036196, August 29, 2025.
45. Musk posted: "This is how great civilizations throughout history have ended. People assume it was due to conquest, but it was actually often simply too much prosperity leading to low birth rate and population collapse, which ultimately enabled them to be conquered." https://x.com/elonmusk/status/1876496976709362045, January 7, 2025.
46. Musk posted: "Humans are weaker and slower than other mammals. If it were not for intelligence, we would have long since died out. Intelligence has obviously been the primary evolutionary vector for humans. What is underappreciated is how much happened in the last 10k years. There was an intelligence explosion ~6k years ago with sudden development of writing, agriculture, elementary math, etc. All that said, our little meat computers are just the biological bootloaders for digital superintelligence." https://x.com/elonmusk/status/1774125951435043021, March 30, 2024.
47. Both are described at length in Bostrom's *Superintelligence* praised by Musk.
48. Musk: "I'm not saying like only smart people should have kids. I'm just saying that smart people should have kids as well. They should at least maintain—at least be a replacement rate. And the fact of the matter is that I notice that a lot of really smart women have zero or

one kid. You're like, 'Wow, that's probably not good.'" Vance, *Elon Musk: Tesla, SpaceX, and the Quest for a Fantastic Future*, 358.

49. Musk: "Gender-affirming care is a terrible euphemism. It's really child sterilization . . . It's child mutilation and sterilization . . . It's incredibly evil and I agree that the people promoting this should go to prison." "Dr. Peterson x Elon Musk," (July 22, 2024), https://x.com/i/broadcasts/1LyGBgPvoDjJN.

Index

Abbott, Greg, 161–2
Accenture, 141
Adams, Douglas, *The Hitchhiker's Guide to the Galaxy* (1979), 10, 35, 132
Adbusters (Canadian magazine), 88–9
AdWords, 32
aerospace industry, USA, 34, 36–7, 39, 40–46, 47–8, 51–3; contracting models, 47, 48–9, 50, 51; rocket production, 41, 42–6, 49–50 *see also* space
Afghanistan, 38, 57, 63
Alphabet, 95, 130
Alternative for Germany party, 128–9
Altman, Sam, 103, 104
Amazon, 60, 121, 130, 141
Andreessen, Marc, 21, 29, 30, 84, 142
Angola, 14–15
"Ani" (pornographic anime companion), 155
anime, Japanese, 13, 107–8, 135, 155
Anti-Defamation League, 132
anti-Semitism, 4, 91–2, 127, 134–5
Apple, 11, 21, 59, 62; iPhone, 70

Arbaugh, Noland, 109
Argentina, 126
artificial intelligence, xi; accounting tools, 145–6; "biological bootloader" idea, 105, 106, 129; Cold War research, 103; DOGE's endpoint as governance by, 147–8, 153, 155; ecological costs of, 132–3, 158, 167; fears about evil robots, 103–4, 106–7, 108, 109, 111, 116–17, 131; generative AI boom (from late-2022), 122, 129–35; and idea of mind control, 110–11; large language models (LLMs), 130–32, 133–8, 145, 151; learning models, 105, 129, 130–32, 133–8, 145, 147, 151; Musk's solution to the risks of, 104, 106–9, 111, 116–17, 122–4, 129, 131–7; Musk's view of future, 163–4; neural networks, 103, 105, 129, 130–32, 133–8; "superintelligent," 103–8, 116–17, 131–2, 133; and Tesla, 58, 78–80, 158; training data/sets, ix, 105, 106, 129, 130, 133 *see also* "mind virus" term

INDEX

Asimov, Isaac, *Foundation* novels, 1, 35
Australia, 76

Barlow, John Perry, "A Declaration of the Independence of Cyberspace" (1996), 18
BASIC programming language, 11
Bastrop, Texas, 161
Battelle, John, 84
Battlestar Galactica (TV show), 12
Belloc, Hilaire, 159
Ben Mhenni, Lina, 89
Bezos, Jeff, 84, 133, 161
Biden, Hunter, 123
Biden, Joe, 54, 61, 115, 117, 136, 149, 150
Bilton, Nick, 96
bitcoin, 94–5, 98
Black Lives Matter, 89–90, 92, 124, 133
Blackwater (private military company), 38, 39, 51
Blade Runner 2049 (film, 2017), 155
Blastar (video game), 11
BMW, 7
Boeing, 47
Booz Allen Hamilton, 38
Bostrom, Nick, 149;
   Superintelligence (2014), 103–4
Botha, Roelof, 25
Brady, Nicholas, 18–19
"Brady Bonds," 18–19
Britain, 92, 141, 159–60
Brown, Michael, 89, 90, 123–4

Brownsville, Texas, 161
Buck Rogers in the 25th Century (TV show), 12
Bukele, Nayib, 129
Burnham, James, 12
Bush, George W., 33, 46, 49, 52, 59, 63, 64

Cambridge Analytica, 92
Canada, 3–4, 15, 17, 18
car manufacturing: Japanese, 43–4, 45, 66, 113; New United Motor Manufacturing, Inc, 66; and trade unions, 66; types of battery, 69–70; in US, 66–7, 68–70, 71–3, 112, 113
Carlson, Tucker, 123, 131
Cebrowski, Arthur, 33, 34, 37, 38–9
The Charlie Rose Show, 68–9, 70
ChatGPT, 129–30, 131, 137
China, 60, 70, 71, 73, 79; electric vehicles in, 73, 74, 112–13; intensifying rivalry with USA, x, 60–61, 73, 112–13; and Tesla, 74, 112–13
Christensen, Clayton, *The Innovator's Dilemma* (1997), 23
Chrysler, 69
CIA, 37, 111
Civilization (video game), 29
climate crisis, 156; and burning of fossil fuels, 57, 69, 75; ecological costs of AI, 132–3, 158, 167; geoengineering technologies, 157–8; Gore's *An Inconvenient Truth*, 62; mitigation and

226

INDEX

adaptation, 75–8; Musk on, 69, 104; Paris Agreement (2016), 72–3

Clinton, Bill, 88

Clinton, Hillary, 91

Collins, Simone and Malcolm, 162

Columbia disaster (2003), 49

Commodore, 10–11, 19

Communism, 5, 110–11

Compaq, 23

computer/digital technology: and apartheid South Africa, 5, 7–8, 27; APIs (Application Programming Interfaces), 141, 146, 164; "Book of Life" identification system in South Africa, 7–8; Commodore 64 (early personal computer), 8; Commodore VIC-20 (early personal computer), 10–11, 19; conventional model of software development ("Waterfall"), 41; early personal computers, 10–11; elite techno-maximalism, xi, 30–31, 81–2; modems, 11, 18, 19; Musk's superset analogy, 30–31, 81, 85, 86, 139; rapid digitalization in present era, 81–3; satellite technology, 19, 22, 33–4, 37, 47, 159; Tech becomes "Big Tech", 60; used by anti-apartheid groups, 8; "walled garden" term, viii, ix, 59; world viewed as code, viii–ix, 140, 147–50, 153 *see also* artificial intelligence; internet; social media; Web 2.0 concept

Confinity, 24

Conservative Political Action Conference (2025), 162

Contemporary Amperex Technology Ltd (CATL), 73, 74–5

Covid-19 pandemic, 61, 77, 79, 82, 100–101, 102, 111, 121; right-wing hostility to public health measures, 111–12, 113–14, 116, 117; and Tesla, 111–14, 117

Cramer, Jim, 60

Crawford, Kate, 158

Cruz, Ted, 139

cryptocurrencies, 94–5, 98–100, 101

Cuba, 14–15

"cybernetic collective" concept, ix; brain–computer interfaces, ix, 11, 14, 82, 107–9, 110–11, 139; and Musk's "cyborg turn," 81–2, 87–8, 104–9, 110–14, 116–20, 136–8; Musk's fear of woke superintelligence, 116–17, 131–2, 133; Musk's solution to the risks of AI, 104, 106–9, 111, 116–17, 122–4, 129, 131–7; and Neuralink, 14, 107–9, 110–11, 165; social media as primary site of symbiosis, 104–6, 107, 114, 122, 129, 132, 135; as vulnerable to infection, 82, 106–7, 111, 114–20, 122–4, 131–2, 133, 135, 136–7, 140

cybernetics, 108, 136, 151–2, 164; cybernetic governance, 140, 141–8, 151, 153, 154, 155; cyberstate concept, 18, 155
cyberpunk, 78, 82, 118
cyberspace, 18, 27
cyborg forms, ix, xi; brain implant technology, 14, 107, 109, 110–11, 139; brain–computer interfaces, ix, 11, 14, 82, 107–9, 110–11, 139; Cybernetics Technology Division of DARPA, 108; and cyberpunk, 78, 82, 118; *Cyborg Musk* future scenario, 162–5; Haraway on politics of, 119, 137–8; human recipients of Neuralink chips, 109; logic of optimization, 103, 140, 154; and the mech, 13, 14, 88, 109, 135, 157; Musk's "cyborg turn," ix, 81–2, 87–8, 104–9, 110–14, 116–20, 136–8; Musk's fear of infection/contamination, 82, 106–7, 111, 114–20, 122–4, 131–2, 133, 135, 136–7, 140; progressive potential of, 119; and transgender people, 118–20

Davies, William, 87
Dawkins, Richard, 123; *The Selfish Gene* (1976), 90; "Viruses of the Mind" (article, 1993), 114
De Klerk, W. A., 5
defense contracting: and neoliberal privatization of power, 32–3, 37–40, 46–7, 48, 51–2, 54–6; "Rumsfeld Doctrine," 32–3, 37, 38, 46–7, 48; and space/satellite technology, 33–4, 35–7, 38–40, 47, 48, 159; and SpaceX, 38–40, 47, 48–9, 159; and War on Terror, 37, 38–9, 48 *see also* Pentagon
Delta Clipper (reusable launch vehicle), 47
Department of Government Efficiency (DOGE), 83, 139–40, 142–8, 151–2, 163–4; access to databases/digital infrastructure, 144–7, 151–2, 154–5; "delete, automate and integrate" playbook, 145–7, 154–5; endpoint as governance by AI, 147–8, 153, 155; Musk leaves (May 2025), 153–5; and state's surveillance capacities, 140, 144–5, 146–8, 154–5
D'Eramo, Marco, 94
Diablo 4 (video game), 143, 144
diversity, equity, inclusion (DEI), 147
DocuSign, 100
Doerr, John, 62
Doge (Shiba Inu dog), 97–8, 142–3
Dogecoin, 98–100, 101
Doohan, James, 12
Dorsey, Jack, 89–90, 91, 115–16
dot-com boom, 19, 21–4, 29, 37, 41–2, 84, 88; meltdown (2000), 27–8

INDEX

eBay, 25, 35
Eberhard, Martin, 61–2
ecological issues, 57, 65, 75–8, 157–8; climate change denial, 69; costs of AI, 132–3, 158, 167 *see also* climate crisis
economics: and Covid-19 pandemic, 100, 113, 121; free-trade consensus, 61, 71, 73; globalization, 20, 43–4, 46, 70–71, 73, 156; inflation crisis (2022), 121; monetary policy after 2007–8 crisis, 60, 75, 85, 87, 121; monopolies, 19, 27, 28–9, 39–40, 52, 54–6, 121, 126; new geoeconomic consensus, 73–5; offshoring and outsourcing by USA, 46, 70–71, 112; protectionism, 5, 61
Edwards, Paul, 5
El Salvador, 129
El Segundo, California, 40–41, 45
electric vehicles (EVs), 58, 59, 61–2, 64, 66, 157; California's ZEV program, 65; Chinese production, 73; lithium-ion batteries, 69–70, 71–3, 74–5; neural networks in, 103; Tesla production in China, 74, 112–13 *see also* Tesla
Elez, Marko, 151
energy markets/industry: battery technology, 58, 59, 69–73, 74–5, 76, 77, 79–80, 157; "cleantech" boom in Silicon Valley, 60, 62–7; cleantech collapse (early 2010s), 67–8; energy autonomy goals, 57–9, 61, 67; and Musk, 58–9, 157; Obama's "Green New Deal," 58, 59, 63–6; oil industry, 57, 59, 61, 63, 65–6, 67; "shale revolution," 67; Tesla Energy, 76; US experiment with green capitalism, 58, 59, 63–6 *see also* renewable energy
European Union, 74–5

Facebook, 60, 85, 86–7, 92, 95
Fairbanks, Eve, 16
far-right politics: 4chan website, 90–91, 95, 98, 132; "Great Replacement" theory, 127; Grok as "MechaHitler," 134–5, 136, 137; and language of violence, 159–60; and memetic warfare, 137; Musk's "digital party," 126–9; Musk's support for, x, 125, 126–9, 137, 150–53, 159–61, 165; nativist warnings over low white fertility, 127, 160–61, 162; neo-Nazis, 125; QAnon adherents, 125; "Take the red pill" phrase, 152–3 *see also* right-wing politics
Farrow, Ronan, 55
Federal Aviation Authority (FAA), 51
financial crisis (2007-8), 58, 63, 64, 69; Great Recession, 59, 60; monetary policy after, 60, 75, 85, 87, 121
Fitzgerald, Joan, 67

INDEX

Five Star Movement, Italy, 126
Floyd, George, 114–15, 116, 118, 120
Ford, Henry, vii, ix–x, 71, 81, 161
Ford Motor Company, 5, 7, 71, 72, 113
Fordism, vii–viii, 7, 27, 43–4, 45, 66
fossil fuels, 57, 65–6, 67, 69, 75; oil industry, 57, 59, 61, 63, 65–6, 67
4chan website, 90–91, 95, 98, 132
fracking, 67
France, 75
Fremont, California, 66–7, 68, 111, 112, 113–14, 117
Friedman, Thomas, 58

GameStop, 101, 102
gaming, online/video, 11, 29, 95, 135, 140, 148–9; and DOGE, 143–4, 153; "non-player characters" (NPCs) concept, 16, 149
Gates, Bill, 24, 148
Gebru, Timnit, 104
Gelb, Stephen, 7
General Motors, 66, 68–9, 113
geoengineering technologies, 157–8
Gerbaudo, Paolo, 125–6
Germany, 74, 75, 126, 128–9
Ghost in the Shell (anime film, 1995), 107–8
Gibbon, Edward, *History of the Decline and Fall of the Roman Empire*, 1
Gibson, William, 21, 87
Gingrich, Newt, 26

globalization, 20, 43–4, 46, 70–71, 73, 156
Godwin, Mike, 135
Godzilla, 13–14
Golden Dome missile defense project, 159
Google, 32, 60, 62, 85–6, 95
Gore, Al, 26, 62
Gottl-Ottlilienfeld, Friedrich von, 27
GPS technology, 19, 22, 33–4
Gracias, Antonio, 46
Griffin, Michael, 47–8, 49, 52
Grok (anti-woke AI), viii, 131–3, 137, 155, 165–6; "reinforcement learning from human feedback" (RLHF), 133–4; right-wing outbursts, 134–5, 136, 137
Gulf War (1991), 34

Haldeman, Joshua (Musk's maternal grandfather), 3–4
#HandsUpDontShoot, 90, 123–4
Hanson, Robin, "How to Live in a Simulation" (article, 2001), 149
Haraway, Donna, "A Cyborg Manifesto" (essay, 1985), 119, 137–8
Hawking, Stephen, 109
Hecht, Gabrielle, 5
Heinlein, Robert A., *Stranger in a Strange Land*, 132
historical change, 1, 2, 81
Hitler, Adolf, 134–5

INDEX

Hobbes, Thomas, *Leviathan* (1651), 157
Hungary, 75

IBM, 5, 7, 141
Immigration and Customs Enforcement (ICE), 151
industrial production/manufacturing: in apartheid South Africa, 5, 6, 7; flexible production networks, 43–4, 45; Japanese "lean" production, 43–4, 45, 66; Musk's "dark factory" ambition, 78–80; Musk's fast, iterative approach to, 41–2, 43–6, 51, 68, 112–13; Musk's "lean Fordism" synthesis, 45–6, 66, 68; Musk's vertical integration, 44–5, 61, 66, 71–4, 78, 81, 112; Toyota Production System, 43–4, 45, 66 *see also* aerospace industry, USA; car manufacturing
inequality: anxiety over cyborg fluidities, 118–20; and cyborg politics, 119, 137–8; the "digital divide," 26–7; Musk's support of hierarchy, 16–17, 119–20, 124, 132–3, 137–8; racial, 4–5, 6–10, 12–17, 26, 27, 124, 132–3; reconfiguring of by internet, 27, 28; and rewards of *technosovereignty*, viii, ix, x–xi; social, 27, 78, 119, 124; Peter Thiel's view of, 28
In-Q-Tel, 37, 48

Instagram, 95
intelligence services, US, 38
international law, x
International Style architecture, 6
internet: and antipathy to bureaucracy, 26; becomes an asset class, 21; chatbots going Nazi, 135; as creation of the state, 19, 20–21, 36–7; disenchantment with, 92–3, 97, 122; dot-com boom, 19, 21–4, 27–8, 29, 37, 41–2, 84, 88; and financial fabulism, 23–4, 86, 94; 4chan website, 90–91, 95, 98, 132; manosphere, 153; and monopolies, 27, 28–9, 39–40, 121, 126; Musk's superset analogy, 30–31, 81, 85, 86, 139; organic analogies, 20; and payments/financial transactions, 24–5, 29; privatization of, 19, 21, 22, 36, 62; profitable businesses in 2000s, 60; rapid growth of in 1990s, 19–20, 21, 32; remade into a marketplace, 87; satellite, 53–4, 55–6, 77; traceroute, 116; trolls/trolling, ix, 90, 91–2, 95–6, 101, 132
Iraq, U.S.-led invasion of (2003), 34, 38, 57, 63
Iron Man 2 (film, 2010), 86
Isaac, Mike, 130
Isaacson, Walter, 46, 54–5, 133, 137
Israel, x, 5
Italy, 126, 127, 129

INDEX

Japan, 66; car manufacturing, 43–4, 45, 66, 113; Doge (Shiba Inu dog), 97–8, 142–3; "lean" production methods, 43–4, 45, 66; lithium-ion cells from, 70; nuclear trauma (1945), 13–14
#JeSuisCharlie, 89
Jobs, Steve, 23, 70
John, Nicholas A., 92
Jones, Phil, 94

Karp, Alexander, *The Technological Republic*, 158–9
Kay, Grace, 133–4
Keller, Bill, 33
Kelly, Kevin, 20, 33
Khan, Lina, 117
Kistler Aerospace, 48–9, 51
Kleeman, Jenny, 110
Kleiner Perkins, 62
Koolhaas, Rem, 79–80
Kouri, Greg, 22
Kratsios, Michael, 142
Krishnan, Sriram, 142
Kupor, Scott, 142
Kurzweil, Ray, *The Singularity Is Near* (2005), 109

labor, 57–8, 66, 156; Biden administration policies, 115, 117; federal workforce in USA, 144–5, 151, 154; Musk's attitude to his workers, 41, 42, 45, 72, 124, 144, 154; Tesla as fiercely anti-union, 68, 72, 117
Le Maire, Bruno, 75
Lead Belly (folk singer), 90
Lee Kuan Yew, 5
Lehman Brothers, 58
Levine, Matt, 99, 102
libertarianism, ix, 25–6, 37
LinkedIn, 95
lithium, 58, 69–70, 74–5, 76, 78
Lockheed Martin, 34, 36–7, 41, 47, 50, 52, 159
Lonsdale, Joe, 161
#LoveWins, 89
Lutz, Bob, 68–9, 70

#MAGA, 92
Malaysia, 53
The Manchurian Candidate (film, 1962), 110–11
manga, Japanese, 13
Manning, Ric, 25
manosphere, 153
Maria, Hurricane (2017), 76
Mars, ix, 35, 40, 41, 43
Martin, Trayvon, 89
Marx, Karl, 106
Maser, James, 47, 50
The Matrix (film, 1999), 110, 152–3
Mazzucato, Mariana, 68
Mckesson, DeRay, 89–90
"mechs," 13–14, 88, 109, 135, 157
media, traditional, 22–3, 30–31, 81, 85–6, 92, 94, 139, 149
Meloni, Giorgia, 129
memes, 82, 87, 90–91, 95–7, 142–3; Doge (Shiba Inu dog), 97–8, 142–3; meme coins/stocks, 87, 98–100, 101–2, 113, 142–3;

Musk's "mind virus" term, 114, 116, 120; Pepe the Frog, 91, 132, 143; Tesla as meme stock, 102, 113

Memphis, 132–3, 158

Meta, 121, 130

#MeToo, 90, 115, 124, 133

Microsoft, 30, 42, 95, 130, 135, 141

migrants: as bugs in the codebase, viii–ix, 140, 149–52; and European ethno-nationalists, 126–7, 128–9, 150–51, 159–60, 165; far right "Great Replacement" theory, 127; far-right demand for forced "remigration," 126–7, 128, 150–51, 159–60, 165; Musk's extreme hostility to, viii–ix, x, 126–7, 128, 140, 149–52, 159–60, 165; Palantir's "ImmigrationOS" contract, 151; racist tweets/memes on sexual violence, 127; and "shadow people" notion, 149–52

military forces: of apartheid South Africa, 14–15; concept of "network-centric warfare," 108, 159; dependence on Musk's technology, viii, 39–40, 52, 54–5; drone operations, 66, 159; Golden Dome missile defense project, 159; "mosaic warfare" concept, 54; and private contractors, 38, 54–6, 159; prosthetics for wounded soldiers, 108; Rumsfeld's "technology revolution," 32–3, 37, 38, 46–7, 48; and satellite technology, 19, 22, 33–4, 37, 47, 159; and Silicon Valley's origins, 20; US presence in Afghanistan and Iraq, 34, 38, 57, 63

"mind virus" term, 82–3, 114, 116, 140; Musk propagates "anti-woke mind virus," 124, 129, 131–7; "woke mind virus," viii, 82–3, 116–17, 118, 120, 122–4, 129, 131–7, 165–6

Mithril Capital, 142

MKUltra programme, 111

Mosaic (web browser), 21, 25

Mueller, George, 48

Musk, Elon, ix–x, xi; attitude to his workers, 41, 42, 45, 72, 124, 144, 154; "become meme" joke, 96–7, 102, 106, 132; childhood in apartheid South Africa, 1, 6, 8–14, 17, 19, 27; daughter Vivian, 118; desire to make humanity "multiplanetary," 35, 43, 137; early career in Silicon Valley, 1, 19, 22–4; fathering of children, 118, 161, 162; and the manosphere, 153; and the "mech," 13, 14, 88, 109, 135, 157; move towards right-wing politics, 82, 116–20, 123–4, 125–9, 131–7, 150–53, 159–62, 165; moves to Canada (1989), 15, 17; moves to Los Angeles (2002), 34–5, 40; moving of markets

INDEX

Musk, Elon – *contn'd.*
by tweets, ix, 94–5, 99, 101, 102; negative personal coverage of, 94; NPC as term of abuse, 16, 149; and Obama's "Green New Deal," 58, 59, 63–4, 66; opposes Covid lockdowns, 111–12, 113–14, 116, 117; popularity of plummets, 153; and reality of climate crisis, 75–8; settles in Silicon Valley (1995), 18, 21; the state as essential for empire of, viii–ix, 18–19, 22, 30, 35–46, 48–56, 59, 63–7, 108–9; support for far right, x, 125, 126–9, 137, 150–53, 159–61, 165; and Thiel's *Zero to One* (2014), 28–9; transphobia of, 118–20, 166; uses language of violence, 159–60; vehemence of anti-immigrant sentiment, viii–ix, x, 126–7, 128, 140, 150–52, 165; view of human brain, 104–5, 114; visions of the apocalypse, 35, 77–80; and warfare/military issues, 54–5; wields chainsaw at conference, 162

Musk, Errol (Musk's father), 8, 9, 10, 11, 13

Musk, Kimbal (Musk's brother), 10, 22, 24, 148–9

Musk, Maye (Musk's mother), 4, 15

Muskism, vii–viii; and anti-Semitism, 127, 134–5; apartheid South Africa as cradle of, 5, 16–17, 19, 61; as based on exclusion, viii, x–xi, 78, 126–9, 140, 148, 149–52; *Carbon Musk* future scenario, 157–8; compound approach to education, 161–2; *Compound Musk* future scenario, 160–62; *Contractor Musk* future scenario, 158–9; *Cyborg Musk* future scenario, 162–5; and "demographic decline," 127, 160–62, 166; entrepreneurship as combat, 29; envisions less human future, ix, 78–80, 81–2, 87–8, 104–9, 110–11, 116–20, 136–8; and European ethno-nationalists, 126–7, 128–9, 150–51, 159–60, 165; and financial fabulism, x, 23–4, 86, 94; imagined future profits/growth, x, 23–4, 94; and the Matrix, 110, 143, 144, 152–3; monetizing of attention alchemy, 87, 94–5, 99, 101, 102, 105–6, 113, 114; racial/ethnic expulsion as central pillar of, viii–ix, x–xi, 126–7, 128, 140, 149–52, 159–60; and regulatory structures, 45, 50, 51, 56; rejection of empathy, viii–ix, 140, 148–9; sovereignty-as-a-service idea, 39–40, 53, 54–6, 61, 77; state symbiosis theme, viii–ix, 19, 22–3, 30, 31, 35–46, 48–56, 59, 63–7, 108–9, 113, 142; *techno-sovereignty* idea, viii, ix, x–xi, 2, 39–40, 54–6, 61, 77, 79, 83, 157, 164–7; and theories of power, 29–31, 38–9, 118–20; and vertical integration, 44–5, 61, 66, 71–4, 78, 81, 112; the world as code, viii–ix, 140, 147–50, 153; zero-based budgeting (ZBB), 145–6

Namibia, 14–15, 25
NASA, viii, 36, 38, 39–40, 45, 48–9, 50–52
Nasdaq, 121
National Review (conservative magazine), 3–4, 12
National Science Foundation, 21
Naval Research Laboratory (USA), 34
Navigation Technologies, 22
neoliberalism, 32–3, 37–40, 46–7, 48–50, 51–6
Netflix, 60
Netherlands, 126–7
Netscape, 21, 22
Neuralink, 14, 107–9, 110–11, 165
neurotechnology, 14, 107, 108–9, 110–11, 139
Nevada, 72, 73, 76, 79–80
New York, 79–80, 89
New Zealand, 126
Newsom, Gavin, 111
Nixon, Richard, 57
Northrop Grumman, 41, 159
nuclear weapons, 156; South African, 5, 6, 14

Obama, Barack, 52, 57, 58, 59, 63, 64–6, 67
Occupy Wall Street, 89, 115, 124, 133
O'Connell, Diarmuid, 63–4
Office for Metropolitan Architecture (OMA), 79–80
Office of Force Transformation (USA), 33, 34, 38–9
oil industry, 57, 59, 61, 63, 65–6, 67
OpenAI, 103, 107, 129–30
Oppenheimer, J. Robert, 96
Optimus (humanoid robot), 12–13, 78, 158
O'Reilly, Tim, 84, 85, 88, 141

Palantir, 37, 147, 151, 155, 158–9, 161
Palmer, Jackson, 101
Path of Exile 2 (video game), 143
PayPal, 24–5, 29–30, 35, 37, 149
Pearson, KeShaun, 133
Peloton, 100, 121
Pentagon, viii, 22, 32–4, 37, 39–40, 45, 47, 50, 51; DARPA, 20–21, 36, 39, 47, 108; Office of Force Transformation, 33, 34, 38–9
Pepe the Frog, 91, 132, 143
Peterson, Jordan, 118
PewDiePie, 95
Phillips, Whitney, 91
Pirate Party, Germany, 126
Poland, 55
Polanyi, Karl, *The Great Transformation* (1944), 156–7
Politico, 128–9

politics, global: algorithmic manipulation concerns, 92; de-globalizing world, x–xi, 5, 46, 61, 73–4, 81; "digital parties," 125–9; end of Cold War, 40, 71, 88; far right nativism on fertility rates, 127, 160–61, 162; liberal international order, x, 16; political hashtags, 89–92, 115, 123–4; Web 2.0 as powerful tool for activism, 88–90 see also far-right politics; right-wing politics

Polytopia (smartphone game), 148–9

Poole, Christopher "moot", 90

Postel, Jon, 26

Pretoria, 6–7, 9–10, 14, 17, 27

Pretoria Boys High School, 9–10

Progressive Federal Party (South Africa), 9

Puerto Rico, 76

racism: anti-Semitism, 4, 91–2, 127, 134–5; Black Lives Matter, 89–90, 92, 124, 133; European ethno-nationalists, 126–7, 128–9, 150–51, 159–60, 165; false "white genocide" claims, 16–17, 134, 137, 161; George Floyd protests, 114–15, 116, 118, 120; Grok's right-wing outbursts, 134–5, 136, 137; of Joshua Haldeman, 4; and Muskism, 126–7, 128, 132–3, 134, 150–51, 159–60; Muskism's goal of a purified community, viii–ix, 126–7, 128, 149–52, 159–60; nativist warnings over low white fertility, 127, 160–61, 162; tweets/memes on sexual violence, 127; and "woke" concept, 89–90, 91, 115–16, 120, 123–4, 129, 133

Radio Shack, 11

Ramaphosa, Cyril, 161

Reagan, Ronald, 47, 48, 119, 159

Reddit, 95, 98, 101

Reeves, Keanu, 152

renewable energy, 57–9, 60, 61, 62–8, 72–3; storage systems/batteries, 58, 59, 157

right-wing politics: "anti-woke" forces, 115–19, 120, 123–4, 125, 129, 131–7; "cyborg conservatism," 119–20; Musk's move towards, 82, 116–20, 123–4, 125–9, 131–7, 150–53, 159–62, 165; pushing rightwards of algorithm, ix, 125, 131–7, 140; transphobia of, 118–20, 166; X Premium superbase, 125, 126, 134, 154 see also far-right politics

Rittenhouse, Kyle, 115

Robinhood (brokerage platform), 100–101

Robinson, Tommy, 159

Robotech (TV show), 12, 13, 14

robotics, 12–13, 58, 78–80, 108, 158 see also artificial intelligence

Rogan, Joe, 105, 114, 123, 129, 131, 150

Roiland, Justin, 95–6

Root, E. Merrill, 3–4
Rossetto, Louis, 25–6
Rubin, Dave, 116
Rumsfeld, Donald, 32–3, 37, 38, 46–7, 50
Russia, x, 54–5, 92

Sacks, David, 25, 142
Salesforce, 60
Salvaggio, Eryk, 146
San Francisco, 20, 84, 89, 91, 107, 123, 133
Sanders, Bernie, 124
Savage, Rachel, 9
Schiller, Dan, 27
Schwartz, Mattathias, 91
science and technology: and Apartheid's architects, 4–5, 7–8, 14, 16, 27; Cold War rationale for Big Science, 36, 37–8; federal funding for, 21, 36–40; neurotechnology, 14, 107, 108–9, 110–11, 139; Polanyi's golem, 156–7; Technocracy movement (Canada), 3–4 *see also* artificial intelligence; computer/digital technology; Silicon Valley
science-fiction, vii, 1, 21, 35; cyberpunk subgenre of, 78, 82, 118
Scotiabank, 18–19
Scott, Howard, 3
Seibt, Naomi, 128
September 11 terror attacks, 33, 141
Shanghai, 74, 112–13
Shifman, Limor, 90–91

Silicon Valley: "Agile" form of programming, 41–2, 44, 46; and Biden administration, 117; "cleantech" boom in 2000s, 60, 62–7; cleantech collapse (early 2010s), 67–8; and Covid economic policies, 100, 113, 121; "disruption" concept, 23; enters "hard tech" era (2022), 130; federal aid to under Obama, 64–5; and financial fabulism, x, 23–4, 86, 94; generative AI boom (from late-2022), 122, 129–35; Lockheed's facility in Sunnyvale, 36–7; monopolies created by, 19, 27, 28–9, 39–40, 121, 126; Musk's early career in, 1, 18, 19, 21, 22–4; O'Reilly's Web 2.0 conferences, 84, 85; origins of in government funding, 20–21, 22, 36–7, 85, 158–9; PayPal Mafia, 25, 29–30, 142; platform business model, 35, 81, 84–8, 103, 106, 121–2; platformization of the state, 140, 141–8; reactionary technocracy principle, 27, 28–9, 154; rising dominance of from 1990s, 26–7, 40, 60, 85, 95; and SpaceX culture, 40–41, 42, 45; the state as customer of, viii, 36, 37–40, 42–6, 47, 48–56, 142; tech downturn (2022), 121–2, 130; technokings of, 29–30, 157; and venture capital, 21–2, 23–4, 28, 60, 62

INDEX

Singapore, 5
smartphones, 60, 81, 92, 120, 157
Snowden, Edward, 38, 141
social media: addictive nature of platforms, 86, 92; attention alchemy, 87, 94–5, 99, 100, 101, 102, 105–6, 113, 114; generation of new preferences by, 128; "going viral" term, 86–7, 93, 114; launches in mid-2000s, 85; Musk's online presence, ix, 86, 93–5, 96–7, 99, 101–2, 105–7, 116–20, 122–3, 126–9; new conglomerates emerge in 2010s, 95; and political activism, 88–90, 115–20, 123–4; political hashtags, 89–92, 115, 123–4; public anxiety over in 2020s, 92–3, 97, 122; pushing rightwards of algorithm, ix, 125, 131–7, 140; right-wing "anti-woke" forces, 115–19, 120, 123–4, 125, 129, 131–7; as sites of cyborg symbiosis, 104–6, 107, 114, 122, 129, 132, 135; trolls/trolling, ix, 90, 91–2, 95–6, 101, 132; and "woke" concept, 89–90, 91, 115–19, 123–4, 129, 132, 133; world of memes and phantasms, 82, 87, 90–91, 95–102, 132 *see also* "mind virus" term; Twitter; X
Solyndra (solar company), 67, 68
Sony, 69–70
South Africa, 6–17, 25, 26; Black population under apartheid, 5, 6, 7, 9, 10, 16, 26; as cradle of Muskism, 5, 16–17, 19, 61; false claims of "white genocide," 16–17, 134, 161; fortress futurism of apartheid era, 4–5, 6, 7–8, 13–14, 15, 16, 27, 61; Musk family emigrates to (1950), 4–5; Musk's childhood in, 1, 6, 8–14, 17, 19, 27
The Sovereign Individual (James Dale Davidson and William Rees-Mogg, 1997), 29
Soviet Union, 14–15; launch of Sputnik, 36
space: Apollo missions, 36, 48; Commercial Orbital Transportation Services (COTS), 49–50, 51; and Federal Communications Commission (FCC), 54; International Space Station (ISS), 49–50; launch market for small satellites, 43, 47, 52–4, 55–6; military use of satellites, 19, 33–4, 37, 47, 159; Musk and Mars, ix, 35, 40, 41, 43; Musk's "Project Starfall," 159; Musk's turn to rocketry in early 2000s, 35–7, 38–46; neoliberal privatization of, 47–8, 51–6; as publicly funded business opportunity, 35–7, 38–40, 47, 48–53, 54–6, 108; Reagan's Strategic Defense Initiative (SDI), 47–8, 119, 159; regulatory structure

INDEX

of industry, 45, 50, 51, 56; retirement of the Space Shuttle, 49; satellite internet services, 53–4, 55–6, 77; Space Race, 36; Starlink satellites, 53–4, 55–6, 77, 165; the state as customer, viii, 36, 37–40, 42–6, 47, 48–56; TacSat program, 34, 38–9 *see also* aerospace industry, USA; SpaceX

SpaceX, viii, x, 12, 34–7, 38–42, 52, 71, 111–12, 161; Dragon capsule, 49–50; Falcon 1 rocket, 42–3, 50, 52; Falcon 9 rocket, 49–50, 53; founding of (2002), 34, 35, 40; as global platform for national projects, 53, 55–6, 61; "Project Starfall," 159; and public funding, 35–7, 38–40, 47, 48–53, 54–6, 66, 108; rocket production, 41, 42–6, 49–50; and SDI veterans, 47–8, 119; starts as military contractor, 38–9, 47, 48, 159; wins COTS contract (2006), 49–50, 51

Srnicek, Nick, 85, 95

St. Clair, Ashley, 162

Star Trek, 12, 143

Starlink, viii, 53–6, 77, 165

Starshield, 54–5

the state/governance: collapsing public trust in institutions, x, xi; as customer of Silicon Valley, viii, 36, 37–40, 42–6, 48–53, 54–6, 142; cybernetic governance, 140, 141–8, 151, 153, 154, 155; direct role of in internet development, 19, 20–21, 36–7; "DOGE AI Deregulation Decision Tool," 147; "Grok for Government" suite, 155; Musk's colonizing metaphor, 31; neoliberal privatization of power, 32–3, 37–40, 46–7, 48–53, 54–6; platformization of, 140, 141–8; pre-DOGE efforts to digitally modernize/integrate, 141–2, 146; as scaffolding for Musk's private gain, viii–ix, 18–19, 22, 30, 35–46, 48–56, 59, 63–7, 108–9, 113, 142; Thiel's conditions for escaping, 29–30, 31 *see also* Department of Government Efficiency (DOGE); United States of America

Strategic Defense Initiative (SDI), 47–8, 119, 159

Tahoe Reno Industrial Center (TRIC), 79–80

Tarpenning, Marc, 61–2

Tate, Andrew, 153

Tavoulareas, Emily, 145

Technocracy movement (Canada), 3–4

techno-utopianism, 25–6, 27, 88

television, 12–14

terrorism: September 11 terror attacks, 33; War on Terror, 33, 37, 38–9, 46, 48, 63, 141

INDEX

Tesla, x; AI and robotics, 12–13, 78–80, 158; Autopilot, 60; and China, 74, 112–13; and Covid-19 pandemic, 111–14, 117; as creature of the state, 63–4, 65, 66, 68, 113; Cybertruck, 16, 77, 78; factors in rise of, 59–61; factory in Fremont, California, 66–7, 68, 111, 112, 113–14, 117; as fiercely anti-union employer, 68, 72, 117; founding of (2003), 61–2; Gigafactories producing batteries, 71–3, 74, 76, 79–80; IPO (2010), 66–7; lithium-ion batteries, 69–70, 71–3, 74, 76; Megapack, 76; as meme stock, 102, 113; Musk as "Technoking," 29, 157; Musk becomes CEO (2008), 62; Musk seeks increased control (late-2005), 163; Musk's geoeconomic strategy, 73–5; neural networks in cars, 103; Powerwall home battery system, 76, 77; price-to-earnings ratio, 113; promise of *electric autonomy*, 58–9, 61, 65, 66, 77, 78–80; the Roadster, 65, 70, 77; sales plummet, 153; self-driving vehicles, 78, 80; and "shadow people" notion, 149–52; state loans/subsidies to, 64, 65, 66, 68; Tesla Energy, 76; and Twitter purchase, 122

Thailand, 70, 95

Thiel, Peter, 24–5, 28–30, 31, 37, 42, 142

Top Gun: Maverick (film, DATE), 118

Toyota, 5, 113; Production System, 43–4, 45, 66

trade unions, 57–8, 66, 68, 72, 117

Transformers (TV show), 12–13, 14, 78

transgender people, 118–20, 166

transhumanism, 109

Trump, Donald: awards contracts to Musk, 54, 159; and China, 60–61, 74; and Covid-19 pandemic, 112, 117; friendship with Musk, 58, 82, 143; instrumentalizes Musk, 154–5; One Big Beautiful Bill (July 2025), 153; presidential election (2016), 91–2; presidential election (2020), 115; second administration, x, 54, 83, 107, 139–40, 142–8, 151–5, 159, 161; suspended from Twitter (2021), 123, 125

Tufekci, Zeynep, 147–8

Tumblr, 98

Tunisia, 89

Twitter: and #MAGA, 92; founding of (2006), 85; "ghost employees" notion, 149; as "the hellsite," 97; influence on public opinion, 124–5; Musk becomes serious user, 93–4; Musk purchases, 82, 107, 122–4, 125, 133, 146–7, 153;

INDEX

Musk realigns to right/far-right, 124–9, 131–7; Musk's monetizing of engagement, 87, 94–5, 99, 101, 102, 105–6, 113, 114; Musk's moving of markets by tweets, ix, 94–5, 99, 101, 102; and progressive politics, 89–90, 91, 115–16, 123–4; Trump suspended from (2021), 123, 125; X Premium superbase, 125, 126, 134, 154; zero-based budgeting (ZBB), 145 *see also* X

Uber, 146
Ukraine, Russian invasion of (2022), x, 54–5
United States of America, x; "Brady Bonds," 18–19; Department of Veteran Affairs, 147; Digital Service (USDS), 141, 142; energy autonomy goal, 57–9, 61, 67; experiment with green capitalism, 58, 59, 63–6; fantasy of instant riches, 101–2; George Floyd protests, 114–15, 116, 118, 120; intensifying rivalry with China, x, 60–61, 73, 112–13; National Security Agency (NSA), 38, 141–2; PATRIOT Act (October 2001), 141–2, 146; presidential election (2016), 91–2; presidential election (2020), 115; presidential election (2024), 150; "Rumsfeld Doctrine," 32–3, 37, 38, 46–7, 48; security state after 9/11, 33, 37–8, 141–2; state's direct role in Silicon Valley's origins, 20–21, 22, 36–7, 85, 158–9; surveillance state in, 141–2, 146–7, 151–2, 154–5
Unsworth, Vernon, 95
USAID, 146, 148

Van der Merwe, André Carl, 15
Vance, J. D., 142
Varoufakis, Yanis, 133
Verwoerd, Hendrik, 12
Vidal, Jacques, 108
Vlaardingerbroek, Eva, 126–7

WallStreetBets (Reddit forum), 101
Warwick, Kevin, 109
Warzel, Charlie, 102
Web 2.0 concept: "architecture of participation," 84, 88, 92, 106, 126, 130; monetizing of user data, 82, 84–6, 87, 94–5, 99, 101, 102, 105–6, 113, 114; O'Reilly's conferences in San Francisco conferences, 84, 85; platform paradigm, 35, 81, 84–8, 103, 106, 121–2; as product of monetary policy, 85, 87, 100, 121; Web 1.0 comparisons, 88
Weidel, Alice, 128
Westly, Steve, 64
WhatsApp, 95
Wiener, Norbert, 136
Wilders, Geert, 126

Wired magazine, 20, 25–6, 33, 84, 145–6, 151
"woke" concept: and Dorsey's Twitter, 89–90, 91, 115–16, 123–4; history of term, 90, 115–16; Musk's fear of woke superintelligence, 116–17, 131–2, 133; right-wing "anti-woke" forces, 115–19, 120, 123–4, 125, 129, 131–7; Trump's war against, 143–4, 154 *see also* "mind virus" term
Wolfe, Gary, 25
Wolfenstein 3D (game), 135
Woolsey, James, 66
Worden, Air Force Brigadier General Pete, 42, 47, 48, 52
World Trade Organization (WTO), 71

X: emergence of, 124, 140; Grok integrated into, viii, 132, 137; Musk's "reply-guys," ix, 125, 126, 134, 142; racism and ethno-nationalism on, 126–7, 128–9, 134, 150; xAI acquires (2025), 132 *see also* Twitter
xAI (Musk's AI company), 131–6, 155, 158
X.com, 24, 29
XPRIZE Foundation, 157

Yarvin, Curtis, 125
YouTube, 85

zero-based budgeting (ZBB), 145–6
Zimmerman, George, 89
Zip2 (Musk's first company), 22–4, 86
Zoom, 100
Zuboff, Shoshana, *The Age of Surveillance Capitalism* (2019), 93
Zuckerberg, Mark, 92, 133
Zwitter, Andrej, 142